Euclid's Army

Preparing Land Forces
for Warfare Today

WILLIAM F. OWEN

Howgate Publishing Limited

First published in 2024 by
Howgate Publishing Limited
Station House
50 North Street
Havant
Hampshire
PO9 1QU
Email: info@howgatepublishing.com
Web: www.howgatepublishing.com

British Library Cataloguing-in-Publication Data
A catalogue record for this book is available from the British Library

ISBN 978-1-912440-64-1 (pbk)
ISBN 978-1-912440-65-8 (hbk)
ISBN 978-1-912440-66-5 (ebk - EPUB)

William F. Owen has asserted his right under the Copyright, Designs and Patents Act, 1988, to be identified as the author of this work.

The views expressed in this book are those of the author and do not necessarily reflect official policy or position.

EUCLID'S ARMY
PREPARING LAND FORCES
FOR WARFARE TODAY

To Roni

CONTENTS

TABLES

FOREWORD

Throughout history, professional preparation for warfighting has been the hallmark of successful land forces when called upon to fight and win battles on the land. In this informative and thought-provoking book, William (Wilf) Owen superbly illustrates the constants and accelerants of land warfare and the choices to optimise existing and future forces, for winning in land warfare.

I first met Wilf in 2013 as the British Army was resetting from a decade of counter-insurgency operations in Iraq and Afghanistan to warfighting. Introduced by our now Chief of the General Staff, Wilf was instantly a voice of vision and military common sense. Along with others across UK Defence including the Royal Marines, we employed Wilf to help challenge, interrogate and develop our reset to warfighting competency. It takes time to reset an army from its last operational framing, in our case counter-insurgency. Little did we realise when we started how much the world would change and how great the need would be to generate capable warfighting forces as key foundations of conventional deterrence.

The core of this book is about the science of war. Much has and will continue to be discussed of the art of war. Serious professionals think about the science of how and with what to fight. The very best of them think of the breadth of military science from logistics to command and control, right across the electro-magnetic spectrum. Wilf Owen is one of these.

To the reader, whether student, professional or amateur, this book should be read with an open, curious and forward-thinking mindset. As Wilf reminds us, the constants of warfare remain: breaking the enemies' will and cohesion; the purpose of war as an extension and return to politics. But the opportunities within military science are many, with choices in structures, capabilities and communications to name but a few. The challenge and prize for any Western nation are to design and develop the optimal integrated defence forces, which are relevant to the nation's security needs, which are affordable, and which can be delivered. As Wilf expertly educates and identifies in this book, the options and opportunities in military science are greater than we realise. They lie beyond technology and range from training, through organisation and structures. Blending, connecting and integrating these choices for enhanced land forces into space, cyber, maritime and air offer even greater opportunities. Too often these choices and opportunities are underexamined, too rushed or poorly informed.

An army that rests on its laurels or previous victories is an army on the precipice of imminent defeat. Maintaining an optimal fighting force in any domain demands constant review and evolution of the optimal structure, capabilities and methods for fighting, which comprise an integrated fighting system. This book unashamedly seeks to lead and inform debate of this critical issue for any nation and its forces.

Major General ZR Stenning OBE
Director Strategy, Strategic Command, British Army

ACKNOWLEDGEMENTS

In some way, I am indebted to everyone I have ever crossed paths with professionally or in the last forty years or more, but some stand out, specifically concerning this work.

Most notably, Lt Col Jim Storr (rtd) set me on the path to do the work from which this book sprang and, more than anyone else, pointed me towards the areas where hard questions needed answering.

The late Colin S. Gray was also instrumental in forming many of the ideas underpinning this work, without which I would still be stuck in the weeds of equipment capability. Likewise, Col. Robert Leonhard's writings were also important.

Considerable thanks must go to Lt Col. Patrick Butler (RTD) for his many weeks of providing "adult supervision" when we worked together as contractors on several Army projects and his considerable insights into warfare based on over 30 years of operational experience. —likewise, Lt Col Matthew Whitchurch (RTD) for his insights and assistance in promoting my work.

Colonel Matt Murphy (RA) (rtd) provided considerable assistance on the fires and targeting aspects of this work.

Maj Gerry Long, Brigadier Ben Barry, Peter Roberts, Col Debi Lomax, Graham Longley Brown, Jez Hermer, Darren Restarick, Christopher Lawrence of the Du Puy Institute and the late David E. Johnston from RAND all contributed perspectives to this work in no order of importance.

Many Israeli Defence Force officers also contributed to this work, most of whom do not wish to be acknowledged for security reasons, but I can thank Eado Hecht and Shmuel Shmuel, both of whom provided important historical and contemporary data. Professor Eitan Shamir, Director of the Begin Sadat Centre of Strategic Studies, also facilitated many important conversations with those in IDF force development. It should also be noted that my fellow Israelis never held back their opinions on where my ideas might differ from theirs. Francis Tusa, the Editor of Defence Analysis and veteran defence writer, contributed considerably to getting data on military equipment costs and navigating the 'voodoo economics' that corrupt such data.

I must also note the contribution of two important Websites, Nicholas Drummond's "UK Land Power" – uklandpower.com and the team at "Think Defence" – www.thinkdefence.co.uk – for archives, perspectives and information, which greatly assisted in much crucial detail which was impossible to gain from other sources. It must also be recognised that many British Army and Royal Marine Officers and NCOs contributed to this work. This includes two Chiefs of General Staff, Army, divisional and brigade Commanders, and many more in Army HQ. They are too numerous to mention by names, but prominently, the various Commanders and Staff of 1 (UK) Armoured Infantry Brigade enabled much of this work considerably. The spectrum of assistance spanned engaging me professionally or taking the time to listen, debate, and correspond so that ideas could be tested against users and user opinion held to the fore. Without serving soldiers, there is simply no customer for my ideas.

Regarding committing those ideas to paper, I will be forever grateful to Kirstin Howgate, who planted the seed that created the book, and lastly, my wife, Roni, who forced me to stop talking and thinking and start writing.

ABBREVIATIONS

AFV	armoured fighting vehicle
APC	armoured personnel carrier
ATGM	anti-tank guided missiles
AVLB	armoured vehicle-launched bridge
BMP	Boyevaya Mashina Pyekhoty, meaning "infantry fighting vehicle"
BTR	Bronetransportyor – Russian for APC
BPT	be prepared to
CBRN	Chemical Biological Radiological Nuclear (an alternative expression to NBC)
CPX	Command Post Exercise (X- Exercise)
DERA	Defence Evaluation and Research Agency (UK)
EM	Electromagnetic
EPLS	Enhanced Palletised Loading System FAE Fuel Air Explosives
FCU	Fuel Consumption Unit
IDF	Israeli Defence Force
IFV	Infantry fighting vehicle
IRB	Improved Ribbon Bridge
ISO	International Shipping Organisation
ISR	Intelligence Surveillance Reconnaissance
ISTAR	Intelligence Surveillance Targeting and Reconnaissance

JLTV	Joint Light Tac-tical Vehicle
JSDF	Japanese Self Defence Force
KIA	Killed in Action
LOAC	Law of Armed Conflict
MBT	Main Battle Tank
MDP	main defensive position
MOBU	memorandum of ballistic understanding
MTBF	mean time between failure
MTI	moving target indicator
NBC	Nuclear Biological Chemical
NCO	non-commissioned officer
NLOS	Non-Line of Sight
OODA	observer, orientate, decide, Act
OP	Observation Post
PACE	primary, alternate, contingency and emergency
POW	Prisoner of War
RIBS	Rigid Hull Inflatable Boats
RIP	relief in place
RPG	Rocket Propelled Grenade
SAM	Surface to Air Missile
SDR	software-defined radios
SEAD	Suppression and Destruction of Enemy Air Defence
SMCS	shoot, move, communicate, sustain
SME	Subject Matter Expert
UAS	Unmanned Air System
UGV	Unmanned Ground Vehicle
USAF	United States Air Force

INTRODUCTION

On the 11th of July 2014, and a decade ago as this is written, Russian Ground Forces used a barrage of 122mm rockets to hit two Ukrainian battalions in assembly areas in Zelenoplillia in the Luhansk Oblast in Ukraine. Over 30 died and more than 100 were wounded. An Orlan-10 UAS detected the targets and directed the strike. Considering "drones" had been finding targets for artillery since 1982 and before, this event should have been nothing more than a tragedy born from stupidity and lack of training. Yet even the US Army chose to call it an insight into the future of war and caused their Capabilities Integration Centre to initiate the Russian New Generation Warfare Study. In 2016, this resulted in the "Russian New Generation Warfare Handbook", but Zelenoplillia had been old wine in old bottles by any objective measure.

This book is about preparing Western Armies for war in the 21st Century. It aims to impart the focus of attention and spending so those armies can protect the secular, democratic way of life. It is not a book about the future of war. It is about preparing for warfare based on things that can be done today with existing equipment rather than fanciful conjectures, novels and academic treatises inherent to those prophesying about "future wars."

Much, if not all, of what follows was born from my involvement in Warfare and Force Development work for the British Army. This work spanned reviewing and "Red Teaming" tactical doctrine publications through field exercises observing Divisional and formation command to concept and equipment development. As much as it was a privilege to do the work, it was also immensely frustrating. This work is born of that frustration in that there needs to be a reversion to first principles, but this book is not about fixing the British Army. It is about Western Armies in general. The reason is simple. Where the problems exist, they are all the same. The one exception is the United States, which has almost unlimited military resources, so it never attracts the same level of analysis and problem-solving most others must engage with. The solutions for America will be distinct from those needed for others.

Most of the problems can be traced to an erosion and corruption of what used to be called Military Thought and Military Science. The distinction between the two is immaterial, but today, both fields are broken compared to what it was in the 19th Century and most of the 20th century. Military Thought and Military Science aimed to better prepare armies for war. Good work is still done, but it is rare. There are also prominent works that are poor. Poor is a value judgement, and I based that opinion on the fact that few modern works assisted in my journey to write this one. If a published work materially assisted me in writing this book, it features in the text or end notes. There is no bibliography.

Clausewitz considered only recent military history relevant to his work, as any attempt to encompass it all was wholly unnecessary for his thesis. I came to similar conclusions, except where points and insights were enduring and relevant. The biggest challenge was the potential issues created by populist narrative versions of history emanating from both World Wars or those used to support various agendas in future war, urban operations and counterinsurgency. Journalism and social media have done immense damage to the coherency and quality of the debate. Realising that terms such as "grey zone" and "asymmetric warfare" are meaningless helped to focus ideas usefully.

Peer-reviewed published work is also rare compared to the ocean of material available on the internet, some of which are exceptionally good and bad, though not in equal measure. Bad prevails. The world is now awash with self-taught military thinkers who tell us the tank is dead or something is transforming warfare. Yet very few of these people know how to write brigade, divisional orders or training programs. They can talk about equipment but not how it fights as part of a wider system or concept. That said, some exceptional work done by men running websites usefully informed some of the ideas contained herein.

In terms of the demise of military thought and military science, a lot of the fault lies with the professional communities themselves. Much can be traced to a failure to distinguish between poor political choices and asking armies to do things they were and always have been inherently not suited to do. Much of the hard-won expertise inherent to planning to fight the Cold War and thus "wars" was washed away by the reforming zeal of those convinced "counterinsurgency" was the "future of War." It would have been funny had the delusion not had such serious consequences for future Western warfighting readiness.

That is not to say that some of my thinking has not been proven wrong over time, but what has changed has changed for a specific reason. The most important was the realisation that all objective measures of military conduct were constrained by time and money. The ideal and real were far removed because of these two factors. How long and how much it takes to train a tank unit to the required standard is more useful than debating what tank is needed or needed at all.

The next material insight was that while observing Division and Brigade Headquarters, it became obvious that how organisations fought and operated could only exist as plans and orders. While this may seem obvious, the key point was that the rapid and timely production of competent orders directly reflected overall competence. The better equipped and organised a formation or division, the more courses of action were available to the planners. It was also clear that unless I could plan and write orders for Divisional and Brigade operations, not only were my credentials as someone who could point out flaws in the plan lacking, but I was not likely to understand how to fight and sustain those formations to the degree where insights were possible. The same was true of training programs. The additional factor was the obvious realisation that unless you can formulate plans and thus generate orders quickly, you simply lack skill and experience. While I had served in a very junior capacity in both regular and reserve Infantry Battalions, with some time in the Intelligence Corps, I certainly never had the leadership ability, force of personality, skill or determination that being a senior officer requires. Still, I could more than hold my own in conversations in the eyes of those who mattered, which in 2017 resulted in further work to enhance divisional manoeuvre. Much of that work informs the book that follows.

The conversations that should have been the most consequential were those that asked what equipment, training, and organisation the British Army should invest in. What constantly undermined or overcomplicated any insights was the latent, often implied, yet hard-wired belief that for the British Army to be adequately prepared, it needed to accurately predict where, when, and why it would fight. Who it would fight would define all else. I strongly disagreed, which provided another reason to write this book.

The other event that spurred me into writing was the February 2022 Russian invasion of Ukraine, or more precisely, the rush to seek lessons where few existed. To date, nothing reported from Ukraine offered any more insight than the analysis dating from 2014 suggested. Likewise, as of August 2024, the ongoing conflict in Gaza revealed little to nothing unexpected, bar that good training saves lives in war, and almost all the urban operations observations closely tracked those previously identified by the British Army. Lessons may be apparent for those unaware of what history and experience has previously taught to the more committed students of warfare.

Anyone who knows me knows I have long held a strong antipathy for Brigadier Richard Simpkin's many works. He died in 1986, having written five books on armoured and modern warfare. A veteran of WW2, he was instrumental in designing and managing the program that delivered the Chieftain Tank in the 1960s, for which he earned an OBE. He was, by any measure, an accomplished soldier. Many of my criticisms of his work still stand, but in writing this book, I have developed a great respect for what he sought to do and how he tried to do it. Simpkin constantly adhered to a practical approach that spoke to his experience and a strong desire to better prepare the West for war against the Soviet Union. However flawed, I believe many of his works were groundbreaking and remain worthy of study. Simpkin was a soldier writing for other soldiers, so he had a slightly annoying conversational style and a near-complete lack of footnotes. This renders his work no less useful. It is a useful style to emulate. This is not an academic paper, so where facts are not sourced, the internet should suffice, and the academic need for the author to show that they are across the literature does not exist. Anyone who challenges any point's validity will not be dissuaded because it has a citation. Just because General X or Y said something does not automatically create insight or evidence.

I am writing for a professional military audience, so very little regard has been paid to conforming to academic standards. What I think and why I think it should be on the page. What I think is not what you should think, but it is probably something you need to consider. This book exists with the expectation of rocks being thrown because some of the conclusions and implications will be uncomfortable for many readers. To quote Richard Simpkin, "Tanks arouse passions. However well an intentional discussion on the priority of tank characteristics seems to be going, one always senses verbal explosions and table thumping just around the corner." We can substitute many other aspects of equipment, training and organisation for tanks.

Thus, the reader should prepare for some heresies against conventional military thinking. Where they exist, these heresies are not born of any desire to attract the mantle of the avant-garde or be a "Disruptor". Disruption is always best avoided, but serious cures for serious disorders should be considered. Conventional military thinking has considerable merit, but when armies fail to prepare adequately for war, it is seldom because of laziness or stupidity. Armies usually fail because of bad ideas and beliefs. Most will never know which ideas and beliefs are the problem because they have never questioned them. Journalists, commentators, and even retired officers tend to focus on equipment problems, yet equipment choices are almost always attached to concepts and doctrines.

Unless there is an agreed-upon overarching idea, most equipment debates are pointless. Any similar debate is ill-informed unless you know how much equipment costs to procure, run and maintain. Where costs are mentioned in this work, that number is always obtained from open-source information, which should be independently checkable by the reader.

I have sought to avoid what I believe is a raft of pointless debates that have corrupted useful discussion. For example, there is no discussion of race, gender or sexuality concerning those who make up an Army because professional standards should render all those discussions irrelevant. If readers think the "cultural change" discussion is somehow important, then culture can only exist as collective ideas and beliefs; it is thus adequately addressed.

It could be alleged that what conclusions exist in this work are just my opinion. That is true, but hopefully, the information and reasoning that lead to that opinion are either explicitly or implicitly obvious. As someone who has argued against the opinion of the senior officer present numerous times, I have no problem holding difficult views. That said, I have exposed those views to many subject matter experts, so conclusions that will cause retired or serving officers and non-commissioned personnel to "throw their teddy bears in the corner" should have been kept to a minimum.

Lastly, why "Euclid's Army?" Euclid could be called the creator of geometry. More importantly, he used basic ideas called axioms to create more advanced proofs or theorems and propositions on which all his work is based. I hope my work loosely mirrors this approach.

1

TO PREPARE AN ARMY FOR WAR

Preparation and Prediction

No work of military thought can prevent any government from committing its armed forces, however well prepared to military action for ill-considered and unachievable military objectives. What military thought can and should do is best prepare any armed force for any conflict regardless of context. This work proposes that "what good looks like", for better or worse, is largely ascertainable and can be objectively discussed, given good data, practical experience and empirical evidence. Much ink has been spilt on the issue of "winning" or "preparing for future Wars", which essentially makes no sense as the problem is not War but Warfare.

Carl von Clausewitz said this:

> To secure the object, we must render the enemy
> powerless, and that, in theory, is the true aim of warfare.
> That aim takes the place of the object, discarding it as
> something *not actually part of war itself* [emphasis added].[1]

Most people are unaware of this quotation or the gravity of the insight. Yet, it is one of the most important Clausewitz ever made. Many scholars, academics, journalists and even soldiers cite Clausewitz's "war is a continuation of politics by other means" without ever having read or understood the critical insight he made to hold warfare distinct from war. War may well be a continuation of policy. In its purest form, "Warfare" is not. This becomes even more critical when connected with Clausewitz's definition of strategy says that strategy is the "use of engagements for the purpose of the war."[2] Thus, engagements must render the enemy powerless, in part or whole, at once or over time. As Hans Delbruck's refinement of Clausewitz concluded, the two choices in strategy are annihilation or attrition, or a combination of both. Regardless of which, the need is to harm the enemy to such a degree that you break his collective will to fight.

Thus, the purpose of an army is to destroy or defeat other armies or armed groups. Armies do not govern, repair, or restore foreign societies. Armies do not fix things. They break them. An army sent to protect sheep would hunt and kill wolves into extinction. It would not worry about the sheep. If an Army is not optimised with that in mind, then all else is for nought. Moreover, it assumes an army is so skilled in combat that it has spare capacity to do other things, not combat-related.

Fighting is a practical skill that requires the knowledge and ability to equip, train and organise to engage and defeat the enemy. Ideally, that means any enemy, anywhere, anytime, for any reason. Seeking to be more specific might render the force less flexible and useful than would otherwise be the case. Whatever the purpose of "the War", the enemy must be rendered powerless. That is as true today as at any time in recorded history.

Armies or land forces must be focused on warfare's true aim or be functionally useless as political instruments. Preparing for "a war" will always be harder than preparing "the war" you have predicted because many suppose you must know why, when, where and against whom you will fight to prepare effectively. This is entirely mistaken for the simple reason that the future is unknowable.

You cannot predict the future of War or Warfare. It is not extreme to suggest that any work proposing the character or nature of future war and warfare will most likely be misleading.

The history and literature of war are well populated by false prophets who sought to tell armies and governments what the future would hold and how it should be prepared for. The French Naval "jeune école" was wrong about the future of naval warfare, as was Douhet about the influence and decisiveness of the bomber. The British government's 1957 Defence White Paper was produced by the brightest and best military minds of the time and was subsequently proven to be mistaken in its vision of the need for manned aircraft. Many technological influences on warfare tend to fall well short of their advertising. As previously stated, warfare is a practical skill. It can only be understood and conducted with existing equipment, training, and existing organisations. To extrapolate in any substantial way into the future is, at best, to risk looking foolish or, at worse, getting people killed.

One of the greatest fallacies associated with "future war" is the idea that the future concerned or even the present is bound up with events and circumstances being "increasingly complex." This has been highly prevalent in UK MOD thinking for nearly 20 years. In August 2021, the UK's integrated operating concept stated that the "strategic context is increasingly complex, dynamic and competitive."[3] This meant the future would be much more complex than today, and that complexity would keep increasing. Something "complex" is harder to understand than something simple. Logically, this means that at some point, the situations faced will exceed or have already exceeded our capacity to understand them. Archibald Wavell and Winston Churchill went from horse-mounted armies with no tanks or aircraft to nuclear warfare within their professional lifetimes, as did thousands of others. No modern commander has ever seen anything like that rate of change. Neither Churchill nor Wavell seemed particularly confused at any point in their careers.

The other vice of the soothsayers who inhabit these debates is their obsession with technological determinism. They believe that the nature of war and warfare is defined and dependent on future or new technology. This leads to the ridiculous idea that making correct technology decisions will be the defining factor in winning the next war. Thus, there are utterances that "drones are transforming warfare" despite "drones" being on the frontline for over 50 years or more. Incremental evolution is not transformation. This is not pedantry. Words matter.

Everything predicted about the future of the US Armed Forces, thought to be true at 08:30 a.m. on September 11th, 2001, was rendered irrelevant by 16:00 that evening. All the arguments and debates about the future size, shape, and needs of the Israeli Defence Force (IDF) on October 6th, 2023, were radically reshaped by the events of October 7th. Nor was this the first time that Israel was forced into such a realisation, as while many cite the supposed strategic surprise of the nearly exactly 50 years earlier 1973 Yom Kippur War, few have ever noted that the outcome of that war saw the IDF double in size. The 2nd Lebanon War of 2006 exposed many of the IDF's shortcomings almost entirely related to budget cuts and flawed campaign design methods.[4]

All of these realisations are more notable given Israel's unique focus on War and Warfare in comparison to almost any other state. Additionally, this is from an Army that sees itself almost continually at war, with the events described as "wars" being notable upticks in violence. Wars that were not predicted or unprepared for are far more common than those that are. The Falklands War was so surprising that the UK had planned some defence cuts, which, had they occurred, would have made the campaign far harder to conduct. Every assumption about British Defence considered true a week before Argentina's April 2nd invasion, were completely irrelevant by April 3rd. The same was true of the 1991 Gulf War, and for the British Army to deploy one Armoured Division of two brigades to Saudi Arabia, the British Army of the Rhine had to be stripped bare of Corps enablers such as tank transporters. Nothing about the structure, training or organisation of the 1st Armoured Division had ever considered it being deployed outside Germany. Likewise, the British Army's commitment to Helmand Province in Afghanistan was completely unpredicted in any policy literature widely extant in 2001.

Unpredicted should not and usually does not mean surprising, as in something unprepared for. You can prepare for events that were not predicted. While the IDF were surprised in 1973 by a war they didn't predict, they were adequately prepared in terms of equipment, training and organisation, albeit all of these would undergo a degree of updating after the war. This was "the war" they had prepared for, though some senior leadership, including the IDF head of intelligence, had assumed and thus predicted it would never happen. Likewise, the fighting in Gaza in 2023 and 2024 had been prepared for in detail, albeit the initiating event was shocking and unforeseen. In 1982 the British Army had the extant organisational capacity and planning staff that could load the best part of an Infantry Division and attached aircraft and helicopters and sail it literally halfway around the world at less than a week's notice regarding the first ships sailing.

Nothing about that was smooth or elegant, but just over two months later, the War was over, and the Falklands were British. This was for a war no one had predicted. While the 1st Armoured Division had never considered being deployed outside Germany, it was more than adequately trained, equipped and organised to address all the missions assigned to it during the 1st Gulf War. Being trained and prepared for War tends to work regardless of the war.

Thus, seeking to predict why, when, where and against whom you will fight has consistently proven less useful than knowing how to fight regardless of the prediction. It can certainly be said that preparing to fight "the war" may be useful for fighting "a war", but there are risks inherent to that approach. Modern military history is replete with tales of Armies that were somehow well trained, equipped, and organised, but for the "wrong war." The "Blitzkrieg" of May 1940 is probably the most overused example, portraying the Wehrmacht as uniquely well-disposed to a new type of warfare, but the underlying thesis is deeply contestable. The Wehrmacht was mostly horse-drawn, with only the minority of Panzer

Divisions even being close to fully mechanised. The only fully mechanised Army in the world in 1940 was the British, who had commenced full-scale mechanisation as early as 1927 and based on experimentation dating from before WW1.[5] The careless historical observation that armies always prepare to fight the last war is thus largely nonsense.

The point here is not to ignite a debate into the history of the literature and lessons of May 1940 but to emphasise that simplistic military history creates bad ideas and, thus, bad military thought and science when preparing for war. All armies evolved and adapted between 1939 and 1945; no nation was universally better prepared for "modern warfare" than the others. Citing "adaptation" as some definable characteristic of successful armies is banal because all warfare demands adaptation. There has never been a successful army that decided to forgo "adaptation." Adaptation is not a choice if you don't want to become extinct.

There are no points for style in warfare. Thus, it is pointless for historians to debate the conduct of the 1944 Normandy Campaign when campaign success was achieved ahead of schedule and with lower casualties than predicted.[6]

Understanding how to fight and operate regardless of circumstances is the sine qua non of warfare. Knowing how to fight in the future does not require the ability to see into the future or guess where, when, and against whom you might fight. There are two simple reasons for this.

First, explicit and well-defined knowledge of how to fight under almost any conditions exists now, today.

Second, fighting and operating can only come into being as an outcome of training. Indeed, direct experience of combat and operations can supersede or enhance training in terms of what soldiers do, but how, what, and why actions provide the baseline to what soldiers do on operations is a direct outcome of training. This may seem obvious and superfluous, except that most literature concerned with future war seems to wish to predict the nature and character of war but not the training that might prepare armies for future war. It may be useful to suggest that future operations will require armies to operate "more dispersed" and to be "more agile", but what training delivers that capability? How do you train to make that happen?

Cost and Size

A dichotomy often raised in debates about the size and shape of modern armies is threat versus capability. As previously outlined, the British Army that trained to fight the Soviet Army in West Germany was an army well

Formations[a]	1973	1982	Difference
Light Infantry Brigades	12	11	−1
Mechanised and Armoured Brigades	20	51	+31
Heavy Mortar Battalions (120-160mm)	23	11	−12
Towed Artillery battalions	5	12	+7
Self-propelled artillery battalions	22	56	+34
Rocket (MLRS) battalions	1	3	+2

Table 1.1 The IDF, 1973-82
[a] Col Boaz Zalmonwicz, IDF – personal communication.

enough prepared to deal with circumstances that fell short of fighting an existential nuclear war. In 1984, the defence budget was 5.4 percent of GDP. "Threat" is powerful and often the only driver most governments will fund. Absent of threat, the concept of "capability" is mostly nebulous and imprecise but would have to be expressed as the most capable land force realisable within a budget. No empirically based rule says a nation must have an Army of X or Y size or that X or Y size is demonstrative of some military power. To this point, history is instructive.

Again, in the 1973 War, Israel considered itself adequately prepared in terms of its budget and the overall size of its army to be successful in any war its enemies might start. What many do not realise is that after the experience of the 1973 War, the IDF set about doubling the size of its field army to safeguard against the attrition seen in the last war.

In 1973, the IDF had 50 tank battalions, 50 infantry battalions and 55 artillery battalions. In 1982, it had 90 tank battalions, 80 infantry and 80, respectively, albeit in total numbers, the artillery tubes were close to 120 unit equivalents.[7]

A more precise data set is shown in Table 1.1.

Such a force structure's financial impact had severe and long-lasting effects on the Israeli economy, which were only resolved in the 21st century. The biggest increase was capital equipment costs, which meant in 1975, defence spending reached >30 percent of GDP. Budgets didn't stabilise below 20 percent until 1984 and dropped below 10 percent until 1994.[8]

The insight here is that your Army is as large as it needs to be for the nation to feel safe. For the IDF, that meant more armour and artillery, with artillery seeing the major proportion of the increase.

What makes other nations feel safe varies widely. In 1982, Israel's population was about 4 million, and 18.3 percent of its GDP was spent on defence. That same year, non-aligned Sweden saw its major threat as the

Soviet Union and had a population of 8.3 million. It spent 2.85 percent of its GDP on defence. In total US dollars, 1982 saw Israel spend $4.5 billion and Sweden $2.98 billion.[9] That same year, South Korea, with a population of 39.3 million, spent 6.24 percent of its GDP, equating to $4.65 billion, broadly like that of Israel. This indicates that a defence budget and a nation's threat perception are essentially arbitrary. Additionally, there is no correlation between population, GDP, and defence spending. The real difference was that Israel knew what a modern war looked like and spent accordingly. They won the war and then doubled the size of their Army to feel safe enough to win the next one. Sweden could have afforded to spend what Israel did on defence but felt safe enough based on what it had. Sweden, Israel and South Korea had broadly similar conscription models, so it seems probable that the force structure adopted by Israel in terms of training, equipment, and organisation would, in general, be appropriate for Sweden's defence needs despite a very different climate, terrain and cultural construct. To be clear, Sweden faced an existential threat in the shape of the Soviet Union. It could have afforded an Army the size of Israel's, but it chose not to and maybe for good reasons. This is why this work will not discuss the overall size of a land force bar the fact that the nation can afford to correctly equip and train the size of the force they have. The budget is also assumed to be limited, requiring the best use of resources within constraints. This is not to deny that armed forces often seek to justify their need for more money by expanding their requirements to justify increased spending.

The size of the Army is not a military problem. It's a political and economic problem. Both subjects lie outside this work's scope and the realities of military thought. There is also a technical problem with quantifying the size of an army. Actual manpower has almost nothing to do with combat power. The only relevant measure of any army size is what part can be deployed or mobilised on operations regarding units, formations, and Divisions. To have real merit, those units, formations, and Divisions must be well-equipped, trained, organised and sustained in high readiness. The Iraqi Army of 1991 was supposedly the fifth-largest army in the world with substantial combat experience. The reality was that it was poorly trained and poorly commanded; it evaporated in a few weeks. It did not know how to fight and had mostly outdated equipment. That means your army cannot exceed the size where you can afford to equip and train adequately for war.

The hard reality is that all armies are constrained by equipment and training, which raises the wide issue of the supposed requirement for "mass." The oft-uttered statement that armies need mass usually defies useful definition beyond the point of saying that an army needs to be larger than it is. Size is relative. "Mass" should be understood in terms of power and persistence rather than overall size. Size will always be limited by budget, and as previously stated, size cannot exceed the resources needed to match mass with effectiveness.

As with the size of an Army, the debate or discussion between conscription and volunteering is not necessarily a subject for military thought. An all-volunteer force will not necessarily have a widely differing approach to operations than a conscripted one, especially in democracies. The need to conscript a force is entirely political and has no basis in military arguments. In terms of a volunteer force, economic, social, and cultural influences will impact who joins the army but not in what numbers. Conscription only exists to gain mass beyond what an Army can attract in terms of financial or social compensation. Well-paid soldiers may not be good soldiers, and money spent on housing, social care, and salaries may not be reflected in an effective, deployable workforce. A conscripted Army might have substantial advantages in recruiting from across the social and educational spectrum, subverting the stereotypes of middle-class officers and working-class enlisted. Indeed, the entire history of the IDF largely stands testament to this, though not entirely. Competent armies do not need to debate volunteers versus conscripts because it is not a military problem. The starkest possible realisation is that the British Army spends a vast proportion of its budget on housing, pensions and pay. In sharp contrast the Israeli Defence spends almost nothing in comparison.

An Army is as big as it needs to be to serve its political purpose within the resources the state can afford. Given those resources, the Army should decide how to realise them regarding equipment, organisation, and training.

Logically, equipment, training, and organisation should be within the control of the army concerned, but they rarely are. This should be a source of concern. Most flawed equipment programs become so because of political, economic, or industrial factors. It is extremely rare, though not impossible, that military judgment alone sets an equipment idea on the path to disaster. The equipment, training, and organisation measures must be situated in a discussion about the best investment balance. All military power is constrained by budget. How that budget is spent is probably the most important question in preparing an army for war. Indeed, it is hard to conceive of an area of military thought more demonstrative of subject matter expert understanding than the ability to discuss, plan, and implement coherently and effectively in the real world how a land force should be equipped, trained, and organised. This level of understanding is most likely to be found within the practitioner and SME community but that does not mean the debate will be immune from normal human organisational needs and agendas. No commander wants to oversee the reduction of their force structure, and many organisations strongly resist change. Much about military thought is faith-based and nonsensical, but that does not mean hard, logical arguments and data are absent, especially when backed up by costs. History clearly shows the damage created by inter-arm and inter-service rivalry for those wanting a bigger budget slice.

Few experienced commanders would argue that a well-trained force with adequate equipment is preferable to a less trained one with excellent equipment. However, measuring equipment capabilities in ways that can be balanced or traded against training is usually far too abstract to allow for objective quantification. That said, for a finite amount of training time, it can usually be assessed that any equipment capability that creates an increased demand for training, in terms of how to use the equipment, will result in some other form of training not being completed. While this strongly indicates the need for simple, coherent equipment plans to reduce the impact on training, it does not make armies immune from wanting to chase the next silver bullet solution, possession of which will impart some reputational advantage, or so some might suppose.

The point that needs emphasis is that while equipment, training, and organisation are three pillars of preparation for military effectiveness, the issue of cost presides over all else.[10] The interrelation between each pillar regarding cost impact is considerably more complex than cursory understanding might indicate, mostly because of human behaviour, not empirically grounded reason, and logic.

The Monash Division

The normal Government and Defence response to most problems is to spend money to solve them. If you want to better prepare armies for warfare in the 21st century, then surely the answer is a massive investment. It may be, but recent history is replete with defence projects that consumed vast amounts of money, meagre, or negative output.[11] What will be suggested here is that overall force development be pursued from a position of constraints and restrictions rather than the traditional requirements and the need to procure "best in class" when "good enough" will be more than adequate.

My experience in force development work as a contractor for the British Armed Forces has spanned from the aftermath of the Royal Marines "Commando 21" restructuring to the Strike Brigade Concept, plus much in between.

To better understand some of the problems I was presented with, I decided to model a deployable Division in as much detail as was feasible within time and resources. This became "Project Monash" regarding how the entire body of work could be described for briefing purposes[12]. It developed through about five distinct iterations. It has been briefed numerous times, including to two and three-star audiences in the UK Army HQ and the Israeli Defence Force, as well as several civilian research agencies.

Monash was largely shaped by my work as a contractor, both "Red Teaming" and contributing to Battle Group, Formation and Division Tactics Manual, as well as a myriad of "urban doctrine." Various discussions with subject matter experts, Staff planning exercises, CPX HQ observations and examination of the extant literature led to the following observations:

1. The ability of a formation or Division to march, disperse, concentrate, sustain, and manoeuvre was closely tied to vehicle numbers, NOT manpower.
2. Lower numbers of vehicles seem to force simpler unit structures, which, in turn, enable useful outcomes in tactics, operations, planning, and command—with little or no reduction in combat power but an overall reduction in friction. The training made the difference.
3. You cannot employ any formation you cannot afford to procure, train, deploy and sustain in combat, regardless of combat power.

Something not apparent at the time, but certainly something lying dormant within the model, was that what was deemed important in terms of vehicle design choices, which were the most visible aspects of capability in terms of what the Government chose to spend money on, were not obvious at all in terms of what mattered in terms of success. Buying "the best" could negatively impact the model. The work commenced in 2015 in discussion with Army HQ in Andover and from direct contact with the Chief of the General Staff at the time. More specifically, I stated that an effective armoured division could have <2,000 vehicles, and while many senior officers disagreed, the direction was "show me." Sadly, the first round of testing conducted by the UK's Defence Science and Technology Laboratory (Dstl) focussed on equipment choices, not the model itself. That said, the conclusions reached by the military observers were overwhelmingly favourable as to what the model demonstrated.[13]

Models are nothing more than approximations of reality, so what counts as an approximation needs to be the starting point. The thesis advanced here is that force structure can only be understood as an output of unit design. Your formations and Divisions must be assembled from a limited number of unit design types covering all the arms and services a formation or division needs. If you are not forced to make decisive and brutal choices about your priorities, how can you see what is important and what is not?

The unit's combat power is not fixed within its resources but rather a sum of all the combat and service support brought to bear. Unit design conducted in isolation of this risks incoherent force structure, which will not fulfil their intended potential. This should become more apparent as we progress.

The "internet" and various forums are oddly awash with unit designs and people attempting to design their ideal infantry battalion, brigade, or even Corps. This is not a new phenomenon. Some are constructed and advocated by retired officers. The problem with almost all extant "fantasy units", and even those well thought out by well-informed sources, is that they have three problems.

The first is that they are mostly unconstrained. They are almost always ideals, not constrained by cost, manpower, platform numbers, weight or logistical considerations. Training budgets are never or rarely mentioned. In a sense, this is understandable. No car enthusiast dreams of the car "he might be able to afford" rather than the one he wants.

Second, these unconstrained models ("train sets") are not often tied to any real tactical doctrine or training programs. Most organisations can be fought and operated by multiple methods, and any set of manpower and equipment can usually be trained to organise and fight in different ways. Very rarely does the train set approach account for this. What gets replicated is standard organisational concepts but with new equipment. By this logic, a 2023 Tank company equipped with the Leopard 2s would fight like a 1967 tank company equipped with M60s or T-55s.

Third, there is usually a less-than-adequate accounting of what the unit carries and how it sustains itself, yet for the unit organisation to be viable, all this needs to be itemised for planning purposes, so how useful is a unit design not expressible in staff data? For example, I have rarely seen a unit design where the amount of fuel held, and its distribution, being described in detail. If you hold 1/3rd of your total fuel bulk in jerry cans, where are those jerry cans carried, and how are they used in fuel distribution?

However, the most critical issue was constraint as in the total number of vehicles. Solid arguments can be made for other forms of constraint, and additional criteria, such as total weight, can be quantified as a number. Manpower is an obvious and well-trodden constraint, but I had already seen it was not part of the actual problem, and it is far less objective than many suggest. Suppose you have 10 x 4-tonne trucks in your unit support echelon (logistics platoon). Do those trucks have one crew or two? The performance difference between a crew of one or two (or even three) is substantial regarding security, navigation, driving hours and sustainment. Doubling the manpower makes those ten trucks more effective. More simply, if manpower is your constraint, some will finesse their structures by putting only six men in the back of an 8-man APC, which is probably not ideal.

Another viable constraint is linear meters, as in the total distance in meters that a unit occupies.[14] Averagely, a 135-vehicle battalion might occupy 675-750 linear meters depending on the actual vehicles and allowances of tie-downs on the vehicle. Smaller vehicles thus mean more vehicles. Factoring in the tie-downs to secure a vehicle onto the vehicle deck means the location of lashing points and tie-down eyes on the vehicle begin to drive the overall number of vehicles.

All in all, the simplest criterion is the total number of vehicles. To force debate and insight, confirmed by over seven years of work, 100 vehicles per unit means the average infantry or armour unit averages between 110 and 145. What counts as a vehicle needs some qualification, especially for vehicles carried on vehicles, trailers, or unmanned ground vehicles. However, the most useful constraint seems to be the number of vehicles requiring permanently assigned drivers on a unit establishment.

Inherent to using that total number of vehicle constraints is the flow-down insight that every seat in every vehicle is associated with a job description, rank, wage, and even camp accommodation. As previously observed, the issue of tactical doctrine or how the unit fights and operates may seem more abstract and intangible. There may be more than one operational concept for one unit design, and unit design may account for more than one type of organisation. The real issue is that the unit must be capable of doing what it is trained or required to do. For example, a unit with an anti-tank platoon containing only six firing posts and 24 carried rounds may not be able to conduct the blocking action the division concepts of operation demands.

Logistic sustainment and its associated modelling are far more complex than I initially imagined, only because of a lack of data or spectrums of data that enable the required judgements. Simple numbers, such as how many litres of water per man per day, are not simple. You can mandate an amount required carried per sub-unit or held in the support echelon based on approximate experience and historical norms. Still, factor in being dispersed across a wide area in high temperatures with little access to water, and those numbers become less reliable. To claim the insight that in hot and arid climates, you need to carry more water does not suddenly create the additional storage space and weight that this requires. If you mandate, the unit must be able to move 800km without refuelling; validated Fuel Consumption Unit figures for each vehicle are required. The means and methods of fuel collection and distribution also need to be considered and assessed regarding how the formation or Division operates. If you mandated 800km of fuel, did you mandate 800km of vehicle spares? If so, what does that weigh, what does it comprise, and where is it carried? A lack of current or valid reference data is, by far, the greatest challenge in force structure modelling. The data is well understood and accounted for in some areas, but that is far from universal.

Nothing written so far should imply or suggest that this is a subject of infinite and, thus, unsolvable complexity. It is not and never has been, but unit design is a resource-constrained activity with multiple impacts outside the arm or service concerned. Unit design cannot exist in isolation from the wider force, and seemingly irrelevant and banal detail can have wider than supposed impacts, as later discussion will show.

To examine this, we should adopt a model that allows the detail to be discussed in terms of what that model may produce in terms of insights as to how to best prepare land forces for future conflict, in real material terms and not abstract discussion and debate.

The Division in Detail

Firstly, why a division? Why not a collection of Brigades within a Field Army Structure? The overwhelming reason was simplicity, as in a division comprising two brigades.[15] Additionally, having been on the Staff of both

1(UK) and 3(UK) Divisions during various CPXs I had some tangential experience with how a division might be fought. This was amassed across five years and seven Divisional CPXs as part of the senior command group.

That said, I had to develop my own Divisional Tactical Doctrine to fit the force structure model.

The Monash Division is a force structure model comprising 20 units of less than 100 vehicles, making a divisional total of less than 2,000 vehicles. This is not a model of what is needed for modern warfare but a structure out of which we need to squeeze the greatest sustained performance and combat power for x amount of money. Doing this has enabled insights which are applicable in the real world.

The original model comprised the following:

- 1898 Vehicles and 7596 men
- 20 Units
- Approximately 90-95 Vehicles per unit
- 1 C3I (1 x Div HQ, 2 x Brigade HQ)
- 7 Manoeuvre Units – Infantry and Armour
- 5 Artillery (2 Guns, 1 MLRS, 1 Long Range Anti-tank, 1 Air Defence)
- 2 Engineer (1 Armoured, 1 River Crossing)
- 2 Recce (1 Ground, 1 UAS/EW)
- 3 Combat Service Support (Medical, Logistics and Equipment Support).

Originally, each unit was required to be able to drive 400km and carry three days of rations and water. The ammunition scales and casualty evacuation figures were all taken from the available staff data.

The structure of a division is far from universal as it concerns understanding, and this is best shown by the fact that the original word "Division" was applied to mean a "dividing of the field army." In the Israeli Defence Force, they use the word "Ugdah," which means to lump or group together. In other words, the Israeli use of the word is the exact opposite to the English (or French). An Israeli division is a group of Brigades allocated to a "Front level" of command. The British Division is the splitting of a Corps. The Monash model is a stand-alone division requiring context-specific support above the divisional command level, which could be Front, Corps or Field Army.

The intervening 7-8 years of discussion, development, wargaming and research altered the Monash structure to a significant degree.

The altered model comprises the following:

- <2,000 Vehicles and <8,500 men
- 21 Units
- Approximately 90-95 Vehicles per unit
- 1 C3I (1 x Div HQ, 2 x Brigade HQ)
- 8 Manoeuvre Units – Infantry and Cavalry
- 6 Artillery (3 guns, 1 MLRS, 1 UAS/EW, 1 Air Defence)
- 2 Engineer (1 Armoured, 1 River Crossing)
- 4 Combat Service Support (2 Logistics, 1 Medical, and 1 Equipment Support).

Why deserves some explanation. The original concept had paid almost no regard to anything but vehicle numbers and logistics but almost nothing towards equipment costs. My mistaken assumption was that small numbers demanded the best possible equipment. As will be shown, this was entirely wrong. I had failed to understand where the balance of investment lay.

The other major conceptual alteration was the decision to ditch all the doctrinal concepts and orthodoxies I had seen developed in the British Army and adopt a vastly simplified concept of operation. Much of an idea rests on how well it can be explained, thus how simple it is. Everybody understood the idea of constraining the number of vehicles. However, further discussion was always hampered by those briefed attempting to force their doctrinal and cultural beliefs onto the model. It took some time for me to realise that this was largely because they thought the model was being advanced as something better than what they were used to and not a method for gaining deep understanding. The model had to be based on means and methods of fighting and operation, which wargaming, simulation, CPXs, staff studies or trials could test. The model, as developed, was not a solution or alternative to existing training, equipment, or organisation, but it could be! How valid were the insights if they couldn't be a credible alternative? To clarify, the operational concept had to be extremely simple to allow for extremely high execution standards. Simple things that are done very well.

The last conceptual development before writing began was to move the Division away from the Cold War considerations of existential defensive warfighting from a garrison in West Germany to a globally relevant combat force able to fight anywhere. 1 BR Corps was expected to fight the Russians for two to four weeks before the war "went nuclear," or diplomacy triumphed. 1 BR Corps had just two to four weeks of ammunition and no plan to regenerate units beyond that point. In no way did it account for the Israeli experience of continuous warfighting with significant and rapid variations in intensity from "routine security" to major combat operations, all within the space of a month and enduring over decades.

It is somewhat strange that the British Army has subsequently managed to discard the experience of the 1970s and 1980s when infantry, armour and even artillery units and others routinely went from training to fight an existential nuclear war at four hours of readiness to patrolling the fields and streets of Northern Ireland in a counter-terrorism campaign. Almost all British officers were equally comfortable with both. Israel has had a broadly similar experience, as in pulsing between warfighting to routine security operations.

Almost every nation on earth could afford to train and equip a variation of the Monash Division. How many Divisions they might need is highly context-dependent, but the intention is to provide a model for discussion and insights, not a single solution to fit all problems.

Conclusion

You cannot predict; thus, you cannot prepare for "the war" your nation will fight. You can train, equip and organise for "a war" that may happen at any time.

The amount of money the government gives you will define the size of your army, and that means the money has to be well spent to train, equip, and organise for "a war."

What follows for the rest of this work will be based on a constrained force development model so that the arguments, debates and insights are tightly framed around logical, achievable and understandable choices.

Endnotes

1 Carl von Clausewitz, *On War*, Chapter 1 Book 1, Howard and Paret edition 1976.
2 Note, "purpose" from the original German Zweck. The Jolles 1943 translation, generally treated as the better work changes this to "object." Both are applicable despite containing separate emphasis.
3 UK MOD Integrated Operating Concept Aug 22. Page 3.
4 This is an entire debate in and of itself, of which there is an extant body of literature, but this is my opinion based on having spoken to many men who were part of the events. An Air Force Colonel who described the IDF's overall pre-2006 direction as being a wish for a police force backed up by a strong air force best summed it up.
5 Harris, JP, *Men, Ideas and Tanks*, Manchester University Press 1995.
6 Again, this can be a point of historical debate. Anyone doubting the validity of this statement should engage with the literature.

7 "War without End" – Lt Col Eado Hecht, IDF .ppt presentation used on numerous British Army Battlefield Studies.

8 "War without End" – Lt Col Eado Hecht, IDF .ppt presentation used on numerous British Army Battlefield Studies.

9 Macrotrends.net retrieved August 2023.

10 To address this problem, the UK MOD has a conceptual framework, TEPIDOIL—Training Equipment Infrastructure Doctrine Organisation Information Logistics. While useful, I think this is no more than a checklist for understanding the totality of equipment or capability choices.

11 The FRES program might be but one example.

12 Why Monash? General John Monash was a First World War Corps Commander of Australian Jewish heritage. The historian AJP Taylor remarked he was the only WW1 General to have creative originality.

13 In 2024 I asked for sight of the Army Reports from that testing to quote for this work but was informed that these remain classified. What was transmittable was that the thesis seemed proven in that "it worked."

14 A liner lane meter is 1m long and 2m wide. Any vehicle that fits with that will have an LLM number roughly equivalent to its total length. A 2.5m wide vehicle will have an LLM number double its length.

15 Some credit needs to go to Lt Col Jim Storr for the Divisional description he provided in "Human Face of War" and his subsequent development of divisional modelling used in "Battlegroup." – Jim Storr, *Battlegroup*, Helion 2021.

2

WAR WITHOUT TANKS

To quote Stalin, no one can fight with empty hands. All forces need equipment, and equipment costs money to buy and employ. Modern military science is dominated by equipment debates, almost excluding issues such as training and organisation. People see equipment as tangible and real, while training and organisation are abstract and hard to quantify for those with little practical experience.

In this chapter, we will examine a set of ideas that tend to dominate equipment discussion and try to move that idea forward in a way that more usefully creates understanding, but first, the following context is necessary.

Between 1989 and 1990 various sources indicate that the Challenger 1 cost about £2 million. Today, based on published contract prices, a brand-new Leopard-2 will cost more than £16 million UKP or $20 million. In 1990, a UK tank crew's average combined annual salary would be around £54,000. Today, that number would be about £140,000. In 1990, the combined annual salary of a tank crew equated to about 2.7 percent of the cost of the vehicle. Today, that ratio is 0.8 percent. Put another way, in 1990, you had to multiply the crew's salary by 37 to equate to the cost of their vehicle. Today, that number is about 114 or an increase of about 208 percent. Are tanks today 200 percent more effective than in 1990? In 1990, a unit of 58 x Challenger 1s cost £116million. Today, that figure would be more than £928 million. Had tank prices risen with inflation, each vehicle would cost about £5.1 million.[1] This essentially asks one simple question: can £928 million be better spent than if it were spent on 58 tanks? It would seem foolish to advocate for retaining 60-70 tonne main battle tanks if you cannot answer that question.

The iconic worth of the tank as a symbol of military power is mostly based on mythology. As we will see, the recurring theme of "the tank is dead" has little to do with "tanks". If preparing an Army for war is primarily focused on equipment, organisation, and training, then a simplistic understanding of equipment almost always leads to a discussion about "tanks." As should now be obvious, tanks cost money, as any equipment does. Discussions and ideas that do not hold the cost of equipment as central to understanding that equipment's value are demonstrably useless in advancing any comprehension of effectiveness. The entire issue of equipment capability needs to be understood from a cost versus effect perspective. The idea that the number of tanks or the centrality of the tank to some metric of military capability is essentially evidence-free in the context of the costs of modern tank ownership. Supposing an Army gets gifted £928 million, would it make sense to spend that money on buying 58 tanks, each costing £16 million or $ 20 million? The rest of this chapter is based on that one question, albeit not solely on tanks. If you cannot arrive at judgements as to what the equipment budget gets spent on and why in an empirically valid and traceable form of evidence, then you essentially surrender the size and shape of an army to opinion and fashion based on a set of shared beliefs and ideas most likely to default to understanding equipment as tank numbers or an overall matrix of capital equipment entirely separated from the realities of communications, logistics and training costs.

So what is a tank, and why does it matter?

Few soldiers should worry about defining what is and is not a tank, so the debates on the internet about tanks, tank destroyers, and assault guns are irrelevant to the important questions that drive force development. This is more important than it may seem. In his winter 1944 Army Group notes, Field Marshall Montgomery opined that.

"A tank is an armoured vehicle designed to carry about firepower; this definition, once understood, simplifies the problem of the employment of armour on the battlefield". [2]

Tanks can be usefully defined and have been for a very long time. Further distinctions are mostly unhelpful. Tanks are essentially direct-fire artillery, most of which can also be capable of indirect fire, though how useful that is might be an area of debate. Until WW2, and for most of that period, many guns mounted on tanks were originally artillery or anti-tank guns. Even initial purpose-built weapons used ammunition natures from artillery pieces or anti-tank guns. Because it is a direct-fire weapon, it requires a degree of protection because it needs to expose itself to employ its weapon. It must also move around the battlefield to apply that fire as and when required. Tanks are direct-fire weapon systems targeting structures, fortifications, and vehicles, which might include other "tanks." Nothing a tank does is unique to it being a tank. Targeting structures and fortifications can be done by other systems, as can killing vehicles and other tanks. The one principal aspect of a tank which qualifies it as unique is its high level of protection from artillery fragmentation and splinters, compared to all other vehicles and some degree of protection to the enemy tank's main armament in the frontal arc. However, immunity is seldom the case. The primary characteristic of the tank is that it can apply direct-fire artillery as in the main gun while being able to manoeuvre less threatened or affected by indirect artillery fire than other platforms. Can artillery fragmentation kill and disable tanks? Yes, it can, and historically it has done so. Artillery fragmentation is extremely lethal to all vehicles lighter than main battle tanks.

Will tanks survive better than almost all other platforms? Yes, they will. Because of that level of protection, tanks are generally also immune to small arms, heavy machine guns and light automatic cannon fire. The immunity against RPGs, ATGMs and other tank rounds is far less relevant in cost and weight than proof against a 152mm shell detonating anywhere beyond 30m. The survival of Israeli armour on the Golan Heights in 1973 was mostly because the Centurion tanks used by the IDF were substantially more tolerant of 152, 130m and 122mm fragmentation than other designs. Had the IDF tank units present been equipped with AMX-13s or Leopard-1s they may have sustained substantially more vehicle losses than was the case. Modern tanks also have highly capable communications and sensor suites, making them useful for tasks beyond direct fire support. This leads to the aphorism that tanks are like dinner jackets, as in when you need one, nothing else will do and that it is better to have a tank and not need it than to need one and not have it.[3] Sadly, these are the same quality of insight that routinely proclaims the tank as dead despite overwhelming evidence saying otherwise on every occasion. If the tank is the best-protected vehicle on the battlefield, anything that threatens the tank with extinction axiomatically makes all and any armoured fighting vehicles extinct. An anti-tank guided weapon is factually just a guided weapon, so if not used just against tanks, it will be used against something else. Maybe a truck or APC. The same is true of loitering munitions or "first-person view drones." (kamikaze drones) If there are no tanks, there will be no more armoured vehicles or equipment of similar size and cost.

The whole thesis of the tank being dead is and always has been provably moronic for the most obvious and glaring reasons that logic makes available. What threatens the future of tanks is cost, not utility or threat. Two of the probably best-protected Battleships in history, the IJN's Yamato and the Musashi, required hundreds of aircraft sorties to sink. Yamato took 12 bombs and seven torpedo hits, while Mushashi needed 17 bombs and

19 torpedoes to flounder. That is more combined hits than it took to sink every Fleet aircraft carrier the IJN lost in WW2. The point here is that the vulnerability inherent to a system does not define the value, or lack thereof, of that system. Thousands of tanks were lost in WW2, and aircraft loss rates far exceeded those of tanks.

To progress, let us just assume that a direct-fire large-calibre weapon on a well-protected chassis (wheeled or tracked) is a good enough reason to have something called a tank. This is the simplest and most provable idea that avoids abstract discussions of "shock action" and "decisive manoeuvre," which are the advertised sole purview of "tanks" versus any other system. The future and present existence of the tank are cost-dependent on what amount of money it takes for other platforms to do what tanks do.

Suppose a 2-tonne armoured vehicle with a mast-mounted sight can generate category-1 grids (6m accuracy) for an indirect fire platform, allowing the destruction of any target detected. Why invest $20 million in a platform that does not do much more? Every target set a tank could previously address from the 1940's to the 1990s can now be destroyed by something far smaller and cheaper. All the tanks' other benefits can be supplied by other armoured vehicles. All that said, there may remain some tasks best suited to what gets called a tank, and these would be counterattacks and exploitation, where moving 40kph while detecting and killing targets is something tanks are uniquely suited to but doing that does not require a 65-70-tonne Leopard 2 or M1A2. It may only need something around 30 tonnes and may not even need many of them in the Divisional order of battle. The weight and cost of equipment needed within the Division, especially in terms of mobility support, river and obstacle crossing, and equipment sustainment assets, are considerably heavier and more expensive than those needed to support what might be termed a light tank at 30 tonnes. It is fair to ask if this is all true, then why has it not been obvious to a wider community for some time? The problem is that the ideas around tanks and armoured vehicles, in general, have remained remarkably consistent and even stagnant since about the late 1940s. At the heart of those ideas lies the "Triangle."

The Flawed Triangle

Suppose you want to understand why ideas have primacy over equipment and that equipment does not always translate the intent of the ideas that lie behind it. In that case, the "armour triangle" is an excellent example of a well-intentioned but deeply flawed conceptual framework which has held back thinking since the turn of the century. The Triangle is expressed as Protection, Mobility and Firepower (or slight variations of those themes), conjecturing that effective tank design is a balance or series of trade-offs between these three criteria. Yet for over 100 years, no effective presentation or body of literature has described how this translates into actual designs, as witnessed by the mostly pointless debates about what was or was not a "good tank." Worse, attempts to compare tanks almost always default to using the triangle as the basis for comparison; witness the vast body of literature attempting to prove how good or bad various allied or axis WW2 tanks were or were not. The lack of insight is because the Triangle has nothing to do with tanks. It was invented to examine the competing characteristics of battleships and cruisers based on three set criteria. Protection was the weight and dimensions of armour. Mobility was speed and endurance, and Firepower was the weight and range of the main gun salvo. The use of the triangle was strictly limited to the design phase and used to compare differing layouts and requirements. It was never, or should not have been, used to compare existing designs with other ships. These facts have never been hidden and are extremely well described in the extant historical warship design literature, especially those dating to the first 40 years of the twentieth century.

It seems likely that JFC Fuller's less-than-perfect understanding of the issues involved led to the migration of the triangle idea to tanks. Fuller conceptualised the triangle as the need for "hitting," "protecting," and "moving" as those things inherent to modern forms of land warfare.[4,5,6,7] While true, it was, at best, simplistic. Mobility for a warship is essentially speed and/or speed versus un-refuelled range. The issue is vastly more complicated for a tank, as are the attendant relationships with the other criteria. So much so that beyond the requirements phase, it is impossible to generate effective comparisons unless the criteria are extremely well defined, such as road speed and range versus the overall weight of armour protection. Thus, the triangle only produces value when it uses one set measurement, but as will be shown when it comes to land vehicles, picking one criterion to measure is completely pointless and mostly misleading. You cannot usefully predict or measure a tank's level of protection just by picking the thickness of the armour on the glacis or turret face. For example, the Panther tank had supposedly superior armour to the T-34/85 on its glacis but noticeably inferior thickness on the sides of the turret, which could more easily be perforated by artillery fragmentation.

Additionally, how is mobility any different from reliability? Why measure power-to-weight ratios when how often the vehicle breaks down is vastly more important? Nonetheless, a detailed examination of the triangle as a checklist does allow for wider insights.

Mobility

Traditional approaches to mobility have talked about "Tactical", "Operational", and "Strategic" in line with supposed "levels of war." While simple and coherent, these descriptions are imprecise, abstract, and far less useful than many assume. Is driving 10km between towns, 500km across a theatre, and 1,000km between continents tactical, operational, and strategic? Given the actual definitions of strategy and tactics, the descriptions make even less sense until defended as a "yes, but you know what I mean" argument responsible for so many bad military concepts. This model also tends to rear its head whenever the track versus wheels debate or discussion reoccurs. An alternative may provide a better description. This is

- Obstacle Crossing
- Road Marching
- Transportability.

Except for obstacle crossing, these descriptions can apply to vehicles, formations, and units. Obstacle crossing is the ability of a vehicle to surmount an obstacle compared to another vehicle or comply with a stated requirement. Fording to a depth of 1m, crossing a 1.5m trench/gap and traversing a 0.5m step are all examples.

Road Marching is the ability of a vehicle or formation to move by road consistently and reliably to whatever distance is demanded.

Transportability is the ability of the vehicle or formation to be carried on or within ships, aircraft, or railways.

This is not an exercise in semantics. Unless concepts have a clear and simple expression which enables clear and simple understanding, friction is being added for no purpose.

Obstacle Crossing

As previously stated, obstacle crossing is the ability to traverse terrain compared to another platform. It is measurable and quantifiable to a useful degree. If obstacle crossing alone is the basis for comparison, tracks will always trump wheels. At its simplest, the vehicle's ability to move across a given terrain without becoming immobilised is a useful concept of obstacle crossing. The one exception to this condition is one mine strike will render a tracked vehicle immobile. In contrast, any wheeled vehicle with more than four wheels should remain mobile to a degree, excepting any other damage.

This favours small light vehicles with low ground pressure, but a larger, main battle tank may perform better when faced with a two-meter ditch or one-meter-high linear obstacle. To account for this, the ability of a vehicle to move a set distance and direction, regardless of the terrain, does create an effective form of comparison. For example, for any given start point, define another point 10 or 20km accessible by a vehicle. The vehicle that can do so by driving the shortest distance is the one with the greatest mobility potential. Given the greatest reasonable variety of terrains, it should be ascertainable as to where the advantage lies. For commanders, obstacle crossing is the criteria that gives them the greatest number of choices regarding how and where to manoeuvre their forces. Something as simple as a vehicle being able to swim a river can have decisive impacts. As such, obstacle crossing cannot be overstated in importance.

Road Marching

Marching is also an almost certain aspect of deployment. A modern unit should aim to be able to move 800km or 500 miles across a good road network in 24 hours, able to repeat that rate of movement for over 2,000 km before having a full 24 hours to rest. This will not always be possible under certain conditions and with some equipment types. Tracked vehicles would find an 800km march day impossible, with 400km and significant maintenance being the best achieved under most conditions. A March Day consists of 10-12 hours of movement, 8 hours of driver sleep, and the remaining time for repair, refuelling, administration and feeding. If you don't do this in training, then for real, it simply will not go to plan. Why 800km? If you can't do that, the enemy will outmarch you in commercial 4x4 pickups, 40-foot container trucks and school buses, which can easily accomplish 800km daily. 800km per day is achievable for BTR-type vehicles and almost all military 4x4s. As of 2017, Canadian Army LAVIII sub-units could routinely march 800km in a day.

Transportability

At its simplest, this means putting a vehicle onto or within another vehicle. Tank transporters are the simplest example and have existed since WW1. One of the great strengths of the Renault FT-17 was that at 6.5 tonnes, unlike British tanks, it could be loaded by French commercial lorries and driven hundreds of kilometres across the theatre.

British CVR-Ts were routinely moved in 20-foot ISO containers during the Cold War. Two M-113s can fit onto a 40-foot ISO flatbed trailer with no modifications. Of course, as AFVs get larger and heavier, this capacity evaporates. A large number of capable light vehicles can be moved inside ISO containers.

The IDF's Merkava IV tanks are transported on two-axle, 16-wheel trailers with a capacity of 77,500kg. Thus, an IDF Tank battalion can move from the Lebanese to the Egyptian border within a day. What has changed over time is that the cost of a tank transport trailer is now substantially less in percentage terms than the overall cost of the tank itself.

Rail movement was the norm for continental armies since the widespread emergence of rail technology in the mid-19th century. By the 1950s, air power and nuclear weapons had substantially altered how the issue of deploying an army by rail was considered since rail networks were inherently vulnerable to interdiction, but that does not alter the fact that literally billions of tonnes of freight is moved by rail every year and moving land forces by rail is highly efficient and cost-effective, given a reasonable commercial model. If it is cost-effective to move forty-foot containers, then it is probably cost-effective to move land forces via the same means.

Because rail networks are inherently vulnerable, moving by rail will tend to be something before the conflict and in support of training. The real implication here is not so much the deployment of land forces by rail as the sustainment of land forces once the shooting starts. Rail is perhaps the ideal, not the real, and any stores or logistics moved by rail will need enough lift to be distributed across the road network. While supposedly obvious, this can be overlooked, and the Russians did in their 2022 invasion of Ukraine.

Air transport is likewise a matter of pure physics. If the vehicle or vehicles can be driven or loaded into an aircraft, flying them globally or inter-theatre distance is easy. The lighter and smaller the vehicle, the more can be lifted in one sortie. Vehicles are routinely described as being "C-130 transportable" or "A-400M transportable." A C-17 can lift a main battle tank, so almost all else is possible, and it has been since the 1960s given large enough aircraft.

The less recognised potential is the use of wide-body cargo airlines such as Boeing 777 and Airbus 330. For example, a Boeing 777 can carry twelve to fourteen 4x4 light patrol vehicles or five JLTVs. Loading vehicles require pallet lift equipment, which can only be delivered to secure airports.

Firepower

Firepower is probably best understood as a platform's ability or potential to destroy or defeat the greatest range of target sets. The more target sets addressable, the greater the firepower value. In terms of firepower, the modern 120 mm-equipped MBT is very hard to beat, both in terms of the round's actual terminal performance and the so-called "stowed kill potential" that is a function of carried ammunition. In simple but not simplistic terms, the anti-tank potential of a platform should not be discounted as a useful metric. That does not discount the immense value of any vehicle with zero anti-tank or even anti-light armour potential because it is designed to carry infantrymen or other specialists, who may or may not carry anti-armour weapons. The lethality of an armoured personnel carrier (APC) is the men it carries. You can give that APC a vehicle-mounted weapon and turn it into an infantry fighting vehicle (IFV). Still, you are almost certainly adding weight and altering how exposed the vehicle needs to be to accomplish its mission; you may need to consider adding protection. Modern technology makes adding lethality to a vehicle or platform comparatively simple. This all comes with training, sustainment, and maintenance costs, which cannot be discounted, however technically feasible they may be. Not often recognised as firepower is a vehicle's ability to sense, detect, and communicate for purposes other than using its armament. Modern tanks are far from the near-blind, limited-visibility vehicles of WW2 or even the 1980s. Today, tanks are more likely to detect the enemy at range than are dismounted infantry and almost certainly have superior situational awareness and communications.

Protection

Israel Tal, father of the Israeli tank program and the man most credited with designing the famous Merkava tanks, inferred in several statements that "protection enables all else."[8] The tank is synonymous with protection, but armour comes at a cost in every respect. The whole protection issue must be more finely balanced than any other criteria in the triangle.

Protection is often discussed in terms of an onion, as in layers of what provides protection. The outer layer is not being seen or detected. The second layer is not being hit, and then if hit, to not be penetrated, but if penetrated, to not be killed. As a conceptual framework, it is coherent, albeit in no way connected to the triangle or coherent with its ideas. This is yet more proof that a battleship design concept does not travel. Not being detected or hit is mostly cost-neutral as a function of overall size or agility. Adaptive camouflage and active protection systems are now mature technologies, albeit not widely adopted, because while not expensive as a percentage cost of the vehicle, retrofitting such systems to existing fleets does demand a noticeable slice of the budget. The real costs come with not being penetrated or killed because the primary mechanism associated with defeating both kinetic and chemical energy threats is mass, and mass comes at a substantial cost spiral both in terms of money and subsystem debt, as in engine power, cooling, suspension systems plus all the attendant support vehicles, and bridging. Mass primarily impacts mobility, as few mobility criteria are not adversely impacted by high mass.

No other aspect of tank and armoured fighting vehicle design is as consequential as protection and its impact on cost, support, and force structure. As the example presented at the beginning of the chapter shows, cost can be prohibitive. As made clear in Chapter One, no discussion about force design can be held agnostic of cost.

Cost, Weight and Complexity

As should now be clear, the "armour triangle" does not provide adequate guidance on what type of armoured vehicles are preferable or what balance of criteria works best or should be favoured. It cannot tell you what tank to design or what tank is better than another. As conceptual frameworks go, it is a list of talking points.

Alternatively, it may be better to design armoured fighting vehicles or any land system around cost, weight, and complexity.

We have already discussed cost extensively at the beginning of this chapter. Most equipment capability debates are pointless without an understanding of cost. "If money were no object", you would simply buy or develop the best money can buy. The best money can buy has rarely translated into the most effective choice for combat, as numerous costly failed programs demonstrate. Money is a real-world constraint and by far the most important. Limiting cost, weight, and complexity enables an

	MBT H	MBT L
Notional Technical reference	Leopard 2A6	CV-90-120
Crew	4	3 (plus seating for 2 dismounts)
Mass	65-tonnes	32-35 tonnes
Gun	Rheinmetall L44 120	RUAG 120mm L50// Cockerill XC-8 120mm
Running gear	Rear sprocket, metal track, torsion bar suspension	Front sprocket, rubber track, torsion bar suspension
Ammunition	120mm fixed round	120mm fixed round
Engine hp	1,500hp	680hp
Power to weight	23hp/tonne	21hp/tonne
Range	500km (1,200 litres)	500km (525 litres)
Notional Fuel Consumption unit (litres per 100km)	240	105
60' Frontal Arc protection maximum in RHaE	740mm KE 1100mm HEAT	100mm KE STANAG 4569 Level 6
Step	1.15m	1m
Trench	3m	2.4m

Table 2.1   MTB-H to MTB-L comparison

axiomatic reduction in cost if weight is kept within limits. This should enable a better understanding of what factors attract costs which do not result in desirable increases in performance.

Weight, as in mass, is a bit more problematic because it depends on what attracts the platform mass. Accepting the Leopard 2 costs presented at the start of the chapter and factoring in that the Polish variant of the Korean K2 costs about $18 million apiece, the Leopard at 65 tonnes is marginally more expensive than the K2 at 56 tonnes.[9] Comparing like-with-like platform cost is strongly correlated with weight. If a 120mm gun equipped with CV-90 costs less than $10-12 million and weighs 28-32 tonnes, we can see the grounds for discussing the investment balance.

The total number of systems and attendant subsystems best defines complexity.

Table 2.1 compares two approaches to current MBT designs based on proven and available technology.

The MBT-H is a 65-tonne main battle tank with a 120mm L44 gun. The MBT-L is a light tank adapted from an IFV chassis, which weighs about 32-35 tonnes and carries a 120mm L50 gun capable of firing all the same ammunition as the H68. All figures used here are notional for discussion and for this discussion, we will assume a requirement that gives them essentially the same fire controls and optronics. These would both be viable platforms today, using mature technology. The MBT-L is less than half the weight but is substantially less well-protected. However, both vehicles are equally likely to survive 155mm artillery fragmentation at 10m, and both would be killed or severely damaged if hit by an AT-14 Kornet ATGM. If both vehicles had the same active protection system, then both would be equally well protected against the threats that can be addressed.

Complexity is harder to gauge, but the MBT-L is a smaller, lighter vehicle, so probably less complex to a degree, but not significantly. The performance of the main armament is likely not significantly different, and both carry nearly the same amount of ammunition with MBT-L carrying only four more rounds (42 v 46). MBT-L uses an autoloader with a theoretical rate of fire of 20 rounds per minute. The limiting criterion, though, is chamber cooling.

The MBT-H is longer, wider and taller, so it has slightly better obstacle crossing, but not significantly. Depending on armour options, both vehicles have approximately the same power-to-weight ratio, giving the same approximate march speeds. The MBT-L uses a rubber band track, so it has less surface impact, is quieter, suffers less from vibration and has better fuel consumption. Norway, Denmark, the Czech Republic, and Holland have all adopted band tracks for major armoured fighting vehicle programs, and Norway has used them in combat in Afghanistan. MBT-H is too heavy to use band-track. Regarding equipment, support band tracks are heavy unitary items, albeit segmented band tracks are now being developed. Band tracks allow large numbers of tracked vehicles to move across the civilian road system with far less risk of widespread damage, thus reducing training constraints.

OK, so what? The traditional view here may be that the gods of War are not on the side of cheap light tanks, so if "money were no object," you would choose MBT-H every time.

The Israelis would certainly agree, but better insights lie in a more nuanced understanding of cost and weight. It is simplistic to suggest that because a tank is substantially cheaper, you can buy more using the reasoning that "quantity has a quality all of its own." This needs to be treated with some caution. If you buy lots of parachutes, there can be no tension between quality and quantity. Clausewitz did make the better observation that "all things being equal, numbers matter most." Still, he employed the qualification that "all things being equal", a critical detail often forgotten by his critics.

The more nuanced perspective is that armed forces have often spent money on the wrong things, leading to extremely expensive equipment that failed to perform as well as expected, given the greater investment. Perfect is the enemy of good.

Crew training is critical to an armoured unit and formation's performance. All things being equal, the lighter vehicle will cost less to run, spares will be cheaper, and less fuel will be burned. That means more training for the same amount of money. You can train more for the same budget with the lighter vehicle. Therefore, the lighter, cheaper vehicle crew should be better trained. The lighter vehicle can access more terrain if soft soil is a factor, cross more bridges, and be more easily transported in or on other vehicles, including trains, ships and aircraft. The MBT-L burns less than half the fuel per 100km as the MBT-H, meaning fewer fuel bowsers per unit are required, and the formations need to hold less fuel and/or conduct less refuelling. However, the overall number of MBTs in a formation does not make this significant above the unit level. The merit of the MBT-H resides almost entirely in the high levels of protection, particularly those of the frontal arc. This is not trivial, but it is worth considering that whatever kills MBT-L may kill MBT-H and vice versa. Being harder to kill is entirely relative. Cost, weight and complexity inform the debate far more usefully than mobility, firepower and protection.

This does not just apply to tanks. We can then consider a different issue of a short-range air defence system such as RBS-70, Stinger or Starstreak. Mounting weapon stations employing these missiles on a 4x4 light utility vehicle is possible. Mounting them on a 10-14 tonne armoured vehicle is just as possible. If that air defence system provides air cover to an armoured battle group, it will need to keep up with the MBTs and IFVs, so a 4x4 vehicle will not suffice. This is obvious, but if you need air defence systems to protect river crossings, assembly areas, HQs or other high-value locations, can you afford the extra expense versus the possibility that the cost savings may allow for procuring more systems, thus providing greater redundancy? The same is true of 120mm mortars, and the same insights will apply. A towed 120mm mortar will be substantially less complex than a stabilised, turntable-mounted 120mm connected to a digital fire control system.

Yet, the range effect on the target will not materially differ given the same tube length and ammunition. The higher cost and complexity will enable a more rapid and accurate response to a call for fire and fewer rounds used to adjust, but given the advent of hand-held fire control computers, that is now less true than it was. A towed 120mm mortar system still has value. What value versus a vehicle-mounted system can be addressed in a discussion informed by the cost, weight, and complexity trade-offs so the towed system may be found wanting.

It is entirely fair to suggest that something cheaper, lighter and less complex will simply not have the capability of a competing system, but this assumes that like is not being compared with like. The debate only works if the system, vehicle, or platforms are essentially capable of doing the same thing or working to the same overall intent, albeit by different means. Holding some capability as entirely non-discretionary or essential voids the debate that cost, weight and complexity should address. This is often a technique requirement writers use to force the technical proposal towards the system they want to procure and provide a reason for an increased budget. Demanding a 155mm howitzer that has a rate of fire not achievable via a manual system ensures a fully automated platform even though the rate of fire may have no material effect on the target.

Cost, weight and complexity can ruthlessly compare a 155mm system, such as the fully automated BAE Archer, with the hand-loaded Nexter Caesar. Performance or capability is important, but that is inherent to comparing 155mm L/52 systems and not trying to compare a 155mm system with a 105mm system. There may be an argument to try and compare 155mm with 105mm. The Falklands War raised such a question when the Commanding Officer of a 155mm FH-70 regiment raised the issue of what a regiment of 155mm could have achieved versus the 105mm present.[10] Given agreement on the comparable effect on target, that debate could be usefully framed around cost, weight, and complexity. As of 2000, the British Army had weights of fire tables that usefully compared those weights of fire in terms of logistic sustainment.

Constraint

Almost every field of human and personal endeavour is dominated by constraints, which are usually financial and/or material. As already noted, it is extremely rare to see any conversation about equipment or training in an organisation start with the idea of having a limited budget. Military culture is awash with epithets about "bean counters" and equipment produced by the lowest bidder, yet no army or armed force has ever been generated without some type of financial or material constraint. How often do you hear of demands for increased defence spending based on some perceived shortfall or threat that can only be addressed by spending more money? Every problem is perceived as a need for more investment. Armies always want more money. There is substantial organisational benefit in understanding how to spend money better or to make better balances of investment, so this is where the path to decisive success is most likely to lie.

The idea of constraint is distinct from cost weight and complexity. It primarily addresses equipment choices that impact the overall force structure.

As has been shown, constraining weight will usually directly impact cost beyond the cost of one system or platform in isolation.

Nothing written so far should imply or suggest that this is a subject of infinite and, thus, unsolvable complexity. Applying a constraint will not magically create the required answer. However, the hard and often unpalatable choices that constraints force can produce difficult-to-find insights and understandings that would not be apparent from an unconstrained model. Having briefed this approach multiple times to both British Army and Israeli Defence Force audiences, I encountered several British officers who pointed out that, in their opinion, the constraint made it "impossible" to create effective capability. They contended that a range of essential capabilities units needed to be effective and could not exist within the constraints we had discussed. Anecdotes are not evidence, but I encountered similar objections in various forms more than once. Initially, I took these objections seriously until further analysis showed that none made sense, given the model's purpose of forcing uncomfortable choices. If the aim was discomfort, why avoid it? By "impossible", they merely meant "uncomfortable" or "unfamiliar". As previously mentioned, the actual complaint was that if it could be proven, the constraints would cease to be a model used to investigate force structure and become a solution. If it became a solution, this would provide evidence to cut the budget. These were real concerns and entirely valid, but the question remained: given the constraint, what were the best choices?

Conclusion

Tanks are not going to become extinct. If they do, it will not be because of anti-tank technology but what it costs an Army to field a heavily protected vehicle versus the actual usefulness gained in the balance of investment. The reader may have thought this chapter was about tanks. While they did get mentioned, the real point was that protection comes at what may be an unsustainable cost. Firepower, protection and mobility tell you nothing about the equipment you need or how much equipment you can afford or invest in commensurate with your constraints on training and organisation. Considering cost, weight, and complexity across all equipment and capabilities is essential.

Endnotes

1 The data that informs this argument comes from multiple sources but should be easily ascertainable to those who wish to research the subject in detail. The exact precision of each figure is not really what the insight is about.
2 General Montgomery, *The Armoured Division in Battle*, 21st Army Group HQ 1944.
3 I was present in the room where this comment was first made regarding its popular reporting in the press, and the context was entirely about justifying Australia's decision to buy M1 Abrams's at a unit price of around $20 million a copy.
4 Fuller had numerous expressions of this idea, almost certainly copied from the Landships Committee. It is somewhat strange that Fuller is still afforded any credibility in this day and age, despite his mostly banal observations, plagiarism and horrific errors of judgement.
5 Fuller, JFC, *The Reformation of War*, Hutchinson 1923.
6 Fuller, JFC, *The Foundation of the Science of War*, Hutchinson 1926.
7 Fuller, JFC, *On Future Warfare*, Sifton Praed 1928.
8 Wright, Patrick, *Tank*, Faber and Faber 2000.
The best English language examination of this which is more a social than a technical commentary on the idea of "tanks."
9 In 2022 Poland purchased 180 Korean K2 tanks for 3.4 billion dollars.
10 Col John Wilson, RA, former editor of the British Army Review. Personal Communication.

3

CONCEPTS, DOCTRINE, INFORMATION AND COMMAND

Warfare is about ideas. Those ideas should primarily express how an armed force should fight and operate. Important though the equipment is, warfare is not about equipment, or rather, equipment without ideas as to its employment and sustainment as an organised force is just lumps of metal standing in a vehicle park. While this may yet again seem obvious to many, the issue of how an army fights and operates gets substantially less public discussion than how big an army is or how well it is equipped. Within armies and professional literature, there is often no shortage of new ideas about fighting and operating, manifesting themselves in many ways, but usually through concepts and doctrine or ideas and teaching to use the literal meaning of both words. Doctrine usually teaches the application of ideas that concepts develop. The historical and contemporary problem is not the existence of concepts and doctrine but that both fields contain good and bad ideas and teaching. Thus, the most important question arising from concepts and doctrine is whether those ideas and teachings translate into how an Army intends to fight, regardless of circumstances. Armies do not pick the wars their governments or circumstances will ultimately involve them.

How to win wars is not something armies should worry about. This may seem a counter-intuitive and dangerous idea, but it is not. Victory is a term which only applies to the winning of battles and engagements. Correctly used the term should never be applied to wars. Wars are fought for policy objectives, as in conditions and behaviours outside of combat that are realisable amongst people in their everyday lives. A war is won when the policy objectives are realised, not when the last battle has been fought. As stated in Chapter One, the purpose of an Army is to render the enemy powerless. If you render the enemy powerless and that does not win the war, then the war should not have been fought. Would defeating the Taliban have turned Afghanistan into a secular pro-western democracy? Who knows? NATO and the US withdrew and ceded the Government to the Taliban, who were in power before the War. Winning every battle was irrelevant, as it had been in Vietnam and Somalia. At the heart of all military ideas should be the limited task of defeating the armed objector to policy and not the policy itself. Without that constraint, there can never be a coherent approach to warfighting and warfare.

So, how does an army inflict defeat on an enemy? The enemy is defeated when they cannot or will not fight. When collective resistance ceases, the battle is over. The condition most likely to cause that is killing, wounding and being captured or the fear of being killed, wounded, or captured. An army that focuses on those things will rarely if ever, be wrong when faced with any type of armed opponent, be they an armoured division, militia, narco-gang, or terrorist organisation. Likewise, a successful army must sustain killing, wounding, and capturing without being subject to the same harmful effects. Creating and sustaining the will to fight is thus paramount above all else.

Thus, conceptual frameworks that provide hard-edged guidance on what activities and behaviours are likely to be successful. An example of this is the Core Functions.

The core functions are Find, Fix, Strike, and Exploit. They are how all things being equal, you beat anyone, anytime, anywhere, and under any conditions. It also holds that your side does not want to be found, fixed, struck, and exploited, so a clear and coherent understanding of how that might happen and how to avoid it needs to be well understood.

The importance of the core functions cannot be overstated; despite this, they have often been mischaracterised and poorly taught. The origins of the Core Functions date to at least 1903 in Ferdinand Foch's "The Principles of War" as a collection of his staff college lectures. The intention behind functions was to create a campaign planning tool for armies to manoeuvre at the theatre level to bring the enemy to battle. Thus, it was odd that in 2010, the British Army's new edition of ADP-Land Ops rebranded the Core Functions as the "Tactical Framework." While they are a framework, they are far more than tactical. They are "strategic" regarding how Foch would have used the word.

As previously stated, Core Functions are essentially a planning tool. They tell you what you must do to bring the enemy to battle under conditions where he lacks freedom of action and can be decisively struck to enable your force to gain a wider advantage.

There is no shortage of similar conceptual frameworks, for example, Shoot, Move, Communicate, Sustain. In some ways, the SMCS framework is akin to the core functions but somewhat less conceptual. In general terms, the SMCS framework is a set of things you need to be able to do in combat. It applies to every command, training, equipment, and organisation level. Often abbreviated to "shoot, move, communicate" as a mantra for minor or small unit tactics, the sustaining element might be assumed to be axiomatic to all else. As with other frameworks, the combined function delivers more than the sum of their parts. Its utility covers everything from formulating training programs to analysing force structures and individual carried loads. Depending on context, each part is inextricably linked with a trade-off or enforced priority. More than anything else, these are things that anyone concerned about combat power should seek to understand and develop.

"Shoot" means more than just individual direct fire. It encompasses finding a target and deciding to engage it. All you do as a soldier and those who support you might be said to serve that one moment when you engage a target. "Shoot" could be described as lethality, and it is useful to understand the lethal potential of any combat element, from individual infantrymen to Army Group. Shooting is nearly useless unless you can move to apply lethal effect, which is generally more effective, giving greater proximity. The closer you are to a target, the more effective and efficient direct fire becomes in general terms. The same is broadly true for indirect fire. Shooting and moving have long been allied in fire and manoeuvre, which is an underpinning idea that lives within this framework. Fire and manoeuvre are symbiotic processes that allow your force to conduct offensive action or mitigate the effects of the same being conducted against them. One makes no sense without the other. This does not negate the idea of using fire to harass or impede enemy activity, but this must be cognisant of resources expended for that purpose.

Move should be distinct from "manoeuvre", which in definitional rather than doctrinal terms just means "to move to a position of advantage." Gaining a positional advantage is critical to operational success. Still, to become manoeuvre, the movement must be understood far above that of individuals or platforms and be seen as something tied to the organisation and composition of groups, battalions, formations and even corps. Communication is as likewise obvious as "shoot" and "move" but is often insufficiently examined. Communication is more than the ability to pass information. In the context of this framework, it also defines the ability command, so it has a significant cognitive element as it concerns the timely and accurate passage of information and instructions. All of what has gone so far is meaningless unless sustained. Sustaining just means the ability to keep shooting, moving, and communicating. On the most basic level, it is logistics, as in ammunition, rations, fuel, and water, but beyond that, it impacts training and organisation. This means that equipment repair and medical training are all part of the subject matter covered by sustainment. The SMCS framework can be applied from checking individual carried loads to Divisional training programs.

Teaching an Army that lethal force requires Precision, Proportion, and Discrimination is another useful conceptual framework. Precision means to strike only that which needs striking. Proportion relates the amount of force used not excessive to the military advantage sought as it must balance the military advantage against the civilian loss. Discrimination means only striking legitimate targets as in distinguishing between military and civilian where and when possible. There is far more to this than space allows, but the point is that a simple and coherent framework exists to impart these ideas to those who will have to apply them.

As previously stated, there are many conceptual frameworks, some obvious and banal. That said, some are wrong and misleading. Chapter Two examined the Protection, Mobility, and Firepower triangle as something that did not afford the insights claimed of it. One of the most famous conceptual frameworks extant today is the "OODA Loop" as in the Observer, Orientate, Decide, Act, which was John Boyd's model of fighter air combat co-opted by his adherents into the Manoeuvre Warfare Handbook. There is considerable debate on whether the OODA loop has any value or provides insight other than a set of steps for a child to cross a road safely.[1] The thesis of the OODA loop is that faster understanding and accurate comprehension of situations compared to an opponent will render you an advantage. The problem is that the statement just made is more useful than the framework itself because it is the idea that observation leads to correct understanding, as in orientation; thus, decisions and actions do not describe how human beings make decisions. No one in the history of war has advocated for slower decision-making, so training subordinates to make fast but good enough decisions is always good. That understanding predates the invention of the OODA loop, and the descriptions and varying graphical representations of the OODA are not coherent with that simple expression.

The debate about conceptual frameworks is essentially about what ones are useful and which are less useful or outright misleading. Subjecting an enemy to shock, surprise, suppression, and isolation will almost always contribute to their defeat if you have trained your force to apply all those effects in a way that is coherent with the means available.

Conceptual Frameworks are not theories. The test of theory is that it explains extant phenomena and enables a degree of prediction. Thus, armies need to be aware of concepts masquerading as theories to sound as though they are more proven and valuable than they are. Much advertised as "Manoeuvre Warfare Theory", does not explain success and failure in warfare any more than balancing Protection, Mobility, and Firepower explains what a good or bad tank is. How do you measure the success of a tank design? How do you measure the success of a theory of warfare when the description provided is not a theory? If poor military history is the body of evidence for your theory, then how likely is the benefit? Neither Basil Liddell-Hart nor John Boyd were not good historians. Nor was JFC Fuller. More importantly, military history evolves based on research and writing. What everyone thought about the battle of Stalingrad in 1980 is not the same as the understanding in 2023. Liddell-Hart and Boyd both claimed to see patterns where none existed. More importantly, it matters not how elegant and correct a theory of war or warfare might be if it cannot be applied to better prepare an army for war.

Command

Assuming we have established the veracity of the contention that warfare is about ideas and looked at the conceptual frameworks as an example, we can move forward with one of the core ideas that inform how armies should equip, train, and organise to fight in the 21st century.

The central premise of what follows is that it is not situational awareness that wins modern battles and engagements. It is the speed of reaction and decision based on a superior understanding and concentration of fires onto relevant targets to enable the manoeuvre of forces and sensors.

Much has been written about kill chains, sensor-to-shooter loops, and reconnaissance strike complexes in future and contemporary war. What follows will suggest that these are inadequate descriptions of the mechanisms required and that what exists is built on a widespread misunderstanding about modern operations. The concept of sensor-to-shooter loops dating back to WW1 mostly depends on wireless communications, albeit the earliest forms of indirect fire used telephone-based communications and even hand-dropped messages from aircraft to adjust fire. The simple fact remains that the idea of being able to kill any target you can detect became a guiding light for artillery and even air forces from the earliest years of the 20th century.

This 100-year journey progressed from officers in balloons with binoculars to unmanned aerial systems communicating directly with long-range missile firing platforms to engage targets at intra-theatre distances. The British had experimented with powered aircraft and airships in field exercises as early as 1910, with the airship able to use wireless telegraphy to report enemy movements over a 100km distance.[2] Being able to fit wireless communications into aircraft was initially far more important than being able to carry bombs since the aircraft's primary role was reconnaissance.

The simplistic yet widely accepted version of modern warfare is that the "kill chain" has become the defining activity of what most land forces should lean towards. A simple Google search will not disabuse anyone of this notion. I aim to prove that this is essentially wrong.

Thus, the proposition is as follows.

> In modern warfare, electronic and air freedom of action is essential. The benefit realised should not just be the detection and destruction of enemy platforms and personnel but the superiority of command and control that enables the consistent and long-term application of an economy of a force superior to that which the enemy can endure.

There is the statement, and so on with the argument.

Air Freedom of Action

Air superiority does not guarantee victory. Vietnam, Iraq and Afghanistan would seem to support that assertion, but that may be simplistic. The equally simple fact is that freedom of action to employ aircraft, helicopters, and UAS gives a land force an overwhelming advantage under almost any conceivable circumstance, which in recent years has all too often been taken for granted by Western forces. The conjecture that air power has shaped modern armies as much as the tank or modern artillery may be a source of historical debate, but it's not wrong. For this discussion, we will assume that Air superiority also means Electromagnetic Spectrum superiority. The ability to employ unmanned systems, be they ground or air, largely relies on the integrity of wireless communications, except truly autonomous systems, which have yet to be understood in an effectively contested environment. Electronic Warfare, like aircraft, has been around for nearly all of the 20th century. The fundamentals are not changing at a rate that defies comprehension. Still, as with aircraft, a competent land force must understand and employ wireless data communication to the best effect possible. That is not the case if the enemy can use wireless-controlled air and ground vehicles to impede your freedom of action.

The current lexicon of the US Air Force (USAF) differentiates superiority and dominance. A discussion about each could and does take entire books to debate, but superiority may not be total. It is important to note that aircraft of all types are generally relatively expensive, fragile and need an expert operation. Thus, today, unlike in WW2, they are likely to be rare assets requiring careful employment.

Given that a land force with freedom of action and the capability to use the air will have the greatest possible advantage over one without, it should thus be obvious that anything degrades or threatens that freedom of action is likewise an extremely valuable capability. The historical effectiveness of the SAM-7 and other MANPADs is generally debated. Still, their advent into Vietnam in 1972 seems to have impacted USAF and South Vietnamese air operations to a serious degree.[3] Despite gaining considerable military success in 1973, the Israeli Air Force's losses were not trivial, and most aircraft fell to ground-based air defence. Freedom of action is not free from risk, and superiority may be relative, local and temporal. Air power can obviously be contested by surface-based air defence and even small-arms fire in the case of helicopters, so air power is inherently a product of being able to kill, defeat and suppress any form of air defence system that might challenge your use of the air. That may still require low-observable radar technology ("stealth") and electronic warfare to achieve the desired effect or mitigate residual capability. As will be seen later, land forces may be an effective part of destroying enemy air defences.

Air freedom of action over land generally bestows three things:

1. The ability to detect, track and thus target the enemy
2. The ability to strike enemy personnel, equipment, and infrastructure
3. The ability to transport men, equipment, stores, and casualties.

If you can employ aircraft which can detect and kill enemy vehicles (armoured or otherwise) the enemy is usually disadvantaged to a catastrophic degree. The enemy cannot mass armour, artillery or even their associated logistics without such a level of signature reduction to render it unusable in the conventional sense. Some slightly less obvious capabilities flow down from the three major ones, such as the ability to communicate and broadcast communications. In doing so, air power is more effective, especially when it comes to controlling unmanned platforms and the use of data. UAS or drones are not doing anything in 2024 that they were not realised as being able to do in 2000. Thus, the security of the EM spectrum is inexorably linked to air power (and space) itself. So far, no element of this should seem surprising, insightful, or novel. History strongly supports the contention that armoured and mechanised forces cannot survive or manoeuvre without credible air defence. The destruction of Syrian armoured forces by Turkey in February 2020 and the near destruction of Armenian forces by Azerbaijan in the 2nd Nagorno-Karabakh War in September later that year gave a clear example of what all well-informed practitioners knew at the time.

Thus, any land force must secure freedom of action by destroying any enemy system that may contest their air or electronic spectrum use. Striking air defence radars and any wireless emitter that can reasonably be detected at range is an extremely important capability for any land force. Land force fires need to be able to detect and target all enemy air defence and electronic systems, so in terms of electronic warfare and balance of investment, the bias should be towards locating and killing emitters rather than intercepting, monitoring, and jamming. Passive detection can generally be accomplished at range with electro-optical sensors. The current state of the art is that electro-optical sensors can detect and provide targeting data at 70km or more from a UAS capable of a 25,000ft service ceiling and a payload of <100kg.

Given that one side has complete or near complete freedom of air action, do their sensors and intelligence allow them to target the things the enemy needs to fight and survive? If so, the aim should be to destroy and degrade enemy capability in preparation for more decisive engagements.

In conceptually simple language, seeking air and spectrum superiority means conducting reconnaissance to find targets, which are transmitted back to platforms that will destroy or attack the detected targets. Given the appropriate means, many moving armoured vehicles are comparatively easy to detect, target, and defeat. Historically, considerable technical investment has been placed on making this possible, primarily based on the lessons from the Normandy Campaign of 1944, where Allied air superiority significantly negatively affected the Germans' ability to manoeuvre and sustain.

The flow-down effect, generally not appreciated as an outcome from the sensor-to-shooter loop, or rather "kill chains," is that the architecture involved can be easily adapted towards a command network that can do far more than simply service targets. This requires three distinct steps to make into reality:

1. To relate target acquisition to manoeuvre as much as it relates to fires
2. To view the activity that moves and sustains fire platforms as enabling force-wide sustainment
3. To use fire networks as the bearers for all C3I data.

What does this mean? In the most fundamental sense, command networks should cease attempting to generate perfect situational awareness on which to base decisions related to fires, manoeuvres, and sustainment based on separate networks and instead understand how to optimise command based on information inherent to employing fires. How you control fires is the same network (network not channel) you use to command, and vice versa.

Supposedly, the unintended consequence of the kill chain is a sensor-soaked or transparent battlefield. Neither is true, but it is entirely logical to exploit the "red picture" inherent to detecting the enemy with wider command decisions, which means such detection does not automatically generate a fire mission to kill a target, except when that target is of such high value as to make it worthwhile as in an air defence or EW system. The staff work and allocation of resources placed on sustaining and moving fire platforms can easily be adapted to the wider force.

The concept of a transparent battlefield is delusional and removed from practitioner experience because believing you can see all the enemy has or vice versa can only provide an entirely false sense of reality since an often-heard criticism of the desire towards perfect situational awareness, which drove most command network development, was that the reality was never even close to perfect situational awareness but was, in fact, "some of the blue and none of the red," as in you "sort of knew where your forces were but had no clue about the enemy". There is an obvious paradoxical quality to this thesis.

A fires lead network almost exactly inverts that picture to "some of the red and none of the blue." Given the capability of modern IP-based self-forming radio networks, the issue of "blue force tracking" is conceptually and technically far simpler than in past times. Passing on your location data is automatic, extremely simple, and only requires a tiny amount of data. The wider point is that if you have the security of bandwidth to conduct unmanned reconnaissance ISR, then you have almost complete freedom of action in the EM spectrum to conduct almost any command function you might wish. This is not to say any attempt or activity may not be disrupted by jamming or attempts to spoof or hijack the UAS, communications, GPS or sensor feeds. However, doing so requires an active emitter, which can usually be detected and thus targeted. Any functioning wireless link between unmanned ground or air platforms confirms that all command functions can proceed unimpeded with enough electronic spectrum and connectivity. The command function should lean strongly towards generating the data traditionally used by kill chains but re-purpose the same data for a command network optimised for fire and manoeuvre relevant to the phase of operations working at the time.

A modern commander should not be concerned with situational awareness or kill chains. Regarding how some language has evolved, the "Kill Web" should functionally be the command network enabling all else. Command networks need to be optimised to find, track and identify enemy targets to be engaged as and when required by the scheme of manoeuvre. The logistic support and sustainment for fire platforms must be leveraged to provide the same planning data as that for the whole force. It is not absurd to suggest that if you can sustain artillery forward, everything else is easy by comparison.

Command in Application

Another flow-down from the focus of kill chains has been an implied reliance on "artificial intelligence" and machine automation. In my experience, over 90 per cent of what is doable and useful is concerned with machine automation rather than some nebulous concept of artificial intelligence. Data processing and decision support algorithms cover what most elements of command and staff work need in any form of technology support. Using shape recognition software to search terabytes of digital air reconnaissance photography hardly qualifies as advanced technology. Indeed, most sensor support aimed at target detection and tracking is mature and well-understood, as are the battle management systems which provide commanders with the information they need to make decisions.

Machine automation and digital mapping can reasonably predict where to place patrols and observation posts to detect enemy movement based on lines of sight and communication planning ranges. With enough doctrinal information, it can even indicate where an enemy force might be on any terrain. However, this entirely depends on correctly understanding the enemy's operation. It won't tell you where the enemy is, but it can suggest where you may want to point your sensors. This technology is reducing the time and manpower needed to conduct the staff work used to support operations. This is a beneficial but unforeseen consequence of digitising the kill chains to make them faster and more reactive, but some caution needs to be applied. The ability for a UAS operator to identify a target and then automatically pass the target data to a firing platform and even command that platform to fire has existed for over 40 years. It was the ultimate expression of what some armies explicitly called digitisation. The flight time for a 155mm shell has not significantly improved for over 50 years. Neither has it for most rockets, though some cruise missiles have become significantly faster, and there are some hypersonic missiles. Counter battery and call-for-fire reaction times are not likely to improve, given that digital systems are already as fast as possible. Indeed, counter-battery fire can be completely automated. Automation is not currently a technical problem. It is a legal and doctrinal problem. Automating counter-battery fire based on radar and digital acoustic systems is not only possible but is comparatively simple and cheap to do. Both radar and acoustic systems can generate automated commands to fire instructions in 1-5 seconds.

However, the thesis suggests many people are chasing the wrong rabbit down the wrong hole regarding digitised kill chains or reconnaissance strike complexes. The sensor-soaked battlefield should not be hoovering up more and more data to be processed. The proliferation of sensors should lead to a massive reduction in processing and better presentation of information to commanders, enabling more timely and decisive action.

There is some insight to be gained by research into insect flight. Research conducted by Rafal Zbikowski at Cranfield in 2006 compellingly argued that one of the primary differences between bird and insect flight was the control mechanism employed. Birds use their eyesight and force feedback sensed by their wings and semi-circular canals to provide stability during flight.[4] Despite the epithet of "bird brain", the bird's brain is a massive energy-consuming organ compared to an insect with a very meagre nervous system. The insect uses hairs on every part of its body. Like trees or reeds in the wind, these hairs can only sense the speed of the air and the direction. This means that the insect can use hundreds of very simple data points to establish exactly which parts of its body and wings are doing relative to each other. Insect eyesight is also substantially less complex than birds. In simple terms, birds sense very few parameters and must process a massive amount. Insects sense hundreds of very simple parameters and process very little.

So far, we have two arguments. The first says that a system optimised to find the enemy will have significant flow-down benefits, including situational awareness derived from a collated set of target locations. The second says that the more sensors you subject the enemy to, the less time and energy a commander should consume to make effective decisions.

Many sensors should lead to quicker and simpler decisions because the situation is more apparent. If the opposite is true, the current system designers have comprehensively misunderstood the requirement. More information should not mean more processing. Command systems should be insect nervous systems, not bird brains.

The idea of an economy of force means you only expend those resources necessary to achieve the desired mission outcome or sustainment of benefit. Any economy of force advantage demands superior information about where the enemy is, his condition and intention. In simple terms, were two near identical forces to engage each other, the force would be better able to find and kill elements of the other before any decisive engagement and would find itself building better odds for any coming close battle. Most readers would be aware of a conceptual framework the British Army refers to as the Geographic Framework, which defines operations as Deep, Close and Rear. Famously, General Rupert Smith, in his 1990 Fighting Instructions of the UK 1st Armoured Division, qualified these at the Divisional level as his responsibility to "Fight the Deep, Resource the Close and Protect the Rear." This was useful guidance and a good description. There is, however, some missing context, which was absent from later British Army doctrine publications, and that was the necessity for the Deep Battle to degrade the enemy to the extent that the close battle is anti-climactic.[5] In modern parlance, "deep battle" has become "shaping operations." This can be debated regarding semantics and doctrine, but the key point that must rise to the surface is that detecting the enemy at range must be leveraged to degrade his preparedness for combat.

It would be entirely fair to challenge the above by saying, "But this is situational awareness." It is, but you obtained that situational awareness by finding targets for fires to kill, and the conduct of your fight is driven by

Category 1	Category 2	Category 3	Category 4	Category 5	Category 6
6m accuracy	15m accuracy	30m accuracy	91m accuracy	305m accuracy	>305m accuracy

Table 3.1   Targeting Standard JP 3-09.3

manoeuvring and sustaining sensors and fire platforms based on that target picture. Situational awareness is merely awareness. It is not a methodology to impose defeat upon the enemy.

Building a network around the entire land forces need to call for accurate and timely fire against positively identified targets should axiomatically create the optimum command system, which should be more than capable of enhancing manoeuvre and sustainment. Being able to execute successfully fire missions in a dynamic and fast-moving battlefield requires the highest level of simple but accurate information. You need to know where the enemy and your forces are. Own forces need to report their location. That is a very small amount of data; a small amount of additional data can provide all the required sustainment and casualty reporting states. The fire network is the command network. Thus, they are one of the same.

So, how would this all be applied in the real world? In simple terms, it means that from Battle Group through formation to Division and above, the commanders at each relevant level of command need to be equipped with the sensors and communications to effectively search their allocated battle space so that the higher level of command can take the laydown of enemy detected and fuse that into a picture the commander can base his decisions on. Every sensor should be able to produce a category 1-3 grid. See Table 3.1.

All competent armies allocate sensors to battlespace. Doing so quickly, as in re-tasking them against a manoeuvring and competent enemy with electronic warfare capability and air defence, will require considerable practice and competence.

This has wider implications for command. Command has long been completely dependent on communication. Today, that communication encompasses targeting data and controlling fires as much as controlling subordinates and receiving their reports. In the aftermath of the 2006 2nd Lebanon War, there was widespread criticism of some IDF officers as being "Plasma Screen Commanders," which made little sense as a complaint and one strongly contested by those concerned. Commanders need relevant, timely and accurate information. Above the sub-unit level, the idea that this will be facilitated by Commanders being shoulder to shoulder with junior NCOs, binoculars in hand, as bullets splash the ground around you, is at best moronic. Unlike the 1980s or even 1990s, the modern commander at the battle group level and above should know considerably more about the overall situation than his subordinates. A correctly performing digital command network should inform the commander of all his sub-units and enable him to speak or communicate with them directly. Supposedly, this brings modern command directly into conflict with the cultural aspirations of decentralised or "mission command," as it is popularly known.

The conjecture that lies at the heart of mission command is that in combat or on operations, higher commanders will not understand the local situation as well as subordinate commanders because they lack proximity and are thus removed from the reality of what the subordinate commander is facing. This then requires a style of command where the higher commander gives broad direction to the junior commanders who use their judgement to accomplish the mission. This requires high levels of training and trust in junior leaders. There thus comes the complaint that digital command, with both the reality and the potential that higher commanders may have sufficient knowledge to direct subordinates in detail (thus detailed command), leads to the "death of mission command," which is viewed as such a great threat to the good running and order of modern armies that anything that threatens it should be dismissed. This may be a slightly extreme characterisation, but the idea of a modern digital command system that informs commanders via target data and automated position reporting is a threat to a cultural artefact called "mission command" is, at best, ill-informed and Luddite.

Nothing about enabling commanders with better information impacts the essential requirement for subordinate commanders and junior leaders at all levels of the highest possible quality who can use their judgement and initiative when required. Training commanders will and should still force them to make quick decisions without ideal or even any information. Hence, nothing about the benefits of digital command threatens the concept of requiring quick-reacting intuitive commanders who will employ the system. This is not just because it makes good sense but because this concept must realise the possibility of communications being denied or degraded to the extent where the higher commander has limited or no communication with his subordinates, as has often been the case for the last 100 years or more.

The communications conceptual framework of PACE, (primary, alternate, contingency, and emergency) can and should be exported to and recognised within command systems. How command and communication are exercised under all four conditions and what makes each condition apparent should be understood and practised.

The most important equipment program in any land force is its communications. Modern armies should possess IP-based software-defined radios (SDR) capable of self-forming ad hoc networking. Sometimes called mobile ad-hoc networks or MANET, modern IP-based combat net radios significantly differ from previous technology. Today, every vehicle-mounted or hand-held (body-worn) communications device is merely a node on a wide area or even a global network, seamlessly producing connectivity as soon as a correctly configured radio becomes part of the network. In conceptual terms, every radio is its own cell phone and cell tower, albeit operating at far lower bandwidths. Such networks are almost impossible to jam, or direction find due to low power, wide bandwidth, and frequency agility. The same capability exists for HF radios, using frequency hopping and automatic link establishment, which should form the basis for alternate command networks. All sub-units HQs and above should or could be HF and/or SATCOM capable.

Modern software-defined communications networks far exceed most armies' communications demands and capacities despite what the industry may want to sell.

There is also a set requirement for HQs to be small, highly mobile, and very low signature. As such:

- Battle Group – 6-8 staff in 2 vehicles.
- Formation/Brigade – 12-16 staff in 4 vehicles
- Division/Task Force – 24-36 staff in 8-10 vehicles

It should be noted that "staff" applies to those concerned with staff work, not the total manpower of the HQ. This very austere number of staff assumes proven machine automation, digital communications, and the training and selection of staff officers and NCOs. Being highly mobile means moving every 2-8 hours while understanding the practical concerns of sleep and feeding.

Each command level should be able to function on no more than 56kps bandwidth, with the option to increase or decrease as and when required. Regarding electronic warfare, HQs should be hard to find and identify as a Division Tactical HQ looks identical to a Company HQ. Division Main may not be transmitting at all but offloading its traffic to close-by sub-units using very low-powered local-area networks.

The Division should take no more than 12 hours from receipt of the Corps order to units crossing the line of departure. This means the Division has 4 hours to write and transmit orders, the Brigade 2 hours 40 minutes, and the Battle Groups 1 hour 20 minutes.[6]

As previously suggested, the modern command could look like this when sufficiently supported by machine automation, which functions as a decision-support tool. It can be said to be reasonably certain that HQs could function with the same staff and timelines with little to no machine automation. The caveat is that you would need very thorough training and selection of staff officers. Not all officers are suited to staff work, especially at the bleeding edge of the high performance that modern operations require.

Information Warfare

To quote the late Colin S. Gray, there is more to war than warfare. War is fought for political purposes. At the heart of political opinion lies ideas, narratives and opinions. Information warfare is an imprecise term encompassing many ideas, concepts, and opinions. What follows here is distinct from what we have already discussed, where commanders use their forces to supply relevant and timely information to inform command decisions.

The modern thesis suggests that in "the war", there is a decisive form of activity outside military action where battles are won and lost in terms of moral authority flowing from a narrative created by aggrieved entities. While true, this is merely politics as it has always existed. It has little to do with training, equipping and organising the land force because it is entirely connected to "the war" the government chose to conduct. The Army's job is to render the enemy powerless, which has nothing to do with social media or cable news networks. You cannot stop people lying or people being ill informed. You cannot prevent the media from taking sides, inventing war crimes to gain ratings, or being overtly hypocritical by ignoring the crimes of others.

The overwhelming majority of social media is based on cell phone technology. Spectrum superiority means controlling all the sensors on the battlefield, so controlling or destroying civilian cell phone networks is necessary to prevent the enemy from using them for whatever purpose. While much about "information warfare" is a rich seam of academic and social debate, men with guns are the decisive factor in the "argument of kings" or anyone else, including street fights. Information warfare cannot win battles and engagements, and it has nothing to do with preparing land forces for combat. It is especially irrelevant against enemies like the Islamic State, who have an entirely different system of values to modern secular democracies, so a video of burning someone alive will gain more worldwide support in some quarters or have more messaging value than a clever editorial in the New York Times. Nothing about being a "social influencer" will protect you from someone sticking a gun in your mouth and telling you to "drop the phone" because the army you focussed your hopes on was "non-lethal." Warfare is about killing and breaking will. Information Warfare is about political opinions mostly consumed on a small screen in a café. They are not the same.

Conclusion

The common operating picture on which all decision-making relies should be an output of a system dedicated to finding and killing targets, but also one with a wider mission to use the consequences of a successful strike to find more targets and manoeuvre sensors, fires and platforms against them. This is the purest meaning of the Core Functions described as Find, Fix, Strike and Exploit regarding what command should optimise a force to connect concepts with training. To keep finding the enemy separate from targeting him is generating unneeded friction into command. Every detected enemy location, vehicle or suspected presence needs to inform a broad and deep targeting picture on which the capacity to offend the enemy can be built.

Endnotes

1 Storr, Jim, *Human Face of War*, Birmingham War Studies 2009.
2 Batten, Simon, *Futile Exercises*, Wolverhampton Military Studies 2018 page 83.
3 AC-130 A shot down, CH-53 and CH-47 all in 1972.
4 Zbikowski's thesis was about modelling insect flight concerning the development of micro-air vehicles. However, his observation about the control loops inherent to these processes has obvious C2 applications.
5 Army Doctrine Publications, Land Operations 2017 Edition.
6 Storr, Jim, *Something Rotten,* Howgate 2022, Page 107.

4

TRAINING APPROACH

The difference between good armies and bad armies is training. All things being equal, that is the single defining criteria of actual military power versus shopping lists of equipment which may or may not be serviceable.

Training is what most armies do, most of the time. Training allows soldiers who have never fought an armed action to defeat enemy soldiers who may have more experience but are less well-trained; at least, that should be possible.

There is no evidence that modern warfare or the supposed future of conflict will require radically different training for modern armies that would be unfamiliar to today's experienced soldiers or even those of the recent past. Warfare tomorrow will be fought using the training of today. Arguably most successful armies know exactly what good training looks like.

Training in the military sense is far more than learning how to do something. Civilians train to drive buses, trains, and aircraft or to be plumbers, electricians, doctors, or accountants. Regarding skills and knowledge, most military training should be and is simple. The one thing differentiating military training from civilian is that military training should be designed to build physical and psychological resiliency and robustness against suffering and deprivation. The old Roman Army adage of sweating in training to bleed less in War is as true today as it was when first spoken.

Training is best understood as something associated with a command level from individual to Corps and even the Field Army. A platoon will have a defined list of things it should be able to do. That will most likely be described in a manual of "Platoon Tactics," but this may not be ideal. Platoon tactics describe how the platoon should fight and operate. That may be less useful and entirely distinct from a manual or pamphlet describing how to train a platoon training to do the things it should be good at. That type of publication would most likely be called an instructor's handbook.

This handbook would itemise the training exercises required to bring a platoon from the recruit stage to deploy on operations. The latter approach is less common but most likely to be more effective. It might be didactic and rules-based, but that in no way should limit its effectiveness or impact. All training should force junior leaders to make fast, effective decisions. There should be no tension between usefully adhering to process and procedure and employing cunning, guile and creativity. If a soldier cannot do that, he needs to find a job where that skill level isn't required. The cornerstone of any training system is that it should enable the development of a high level of skill by relentlessly focussing on getting very basic things done very well. That overarching concept extends from individual marksmanship to conducting a brigade-level beach landing or getting a division across a 200m wide river. To this end, the hierarchy of training is essential in that organisations cannot train themselves. Every command level must be tested and satisfy its higher headquarters to do exactly what may be demanded. If decisions made by subordinate HQs are not subject to critical evaluation, then there can be no hope for improvement.

All training consumes time and money; therefore, in a resource-constrained environment, there is often a real incentive to reduce training or not supply sufficient resources for it to be done well.

How much training is enough? In real terms, training is about measurable performance, which can be demonstrated and tested. Enough training is the amount that means you can deploy on warfighting operations tomorrow with no additional training.

Something not rehearsed and learned in training will not be done in contact with the enemy.

Individual Training

Training civilians to be soldiers is a well-trodden path, but the product varies depending on culture, nation, and expectations. The national education system is often more relevant than many suppose as is the motivation to join an all-volunteer force. A conscript force tends to be agnostic of that as service is a legal responsibility, so an infantry platoon, tank crew or catering detachment should be a true cross-section of that society. That said, an Army trains at the rate of slowest or, more often, the mean of soldiers' capabilities. Having a few extremely motivated and intelligent soldiers does not automatically raise all those around them. The methods for accounting for varying levels of education and motivation do not vary much between volunteer and conscript armies. Almost all armies have pre-recruit testing standards, either barring entry or streaming recruits into training appropriate to their limitations. Conscript armies do the same, albeit for a far higher number of individuals.

The deciding factor in individual or basic training is how effective the soldier passing out of training is compared to the time and cost it took to get them to that standard. This means pass or fail testing to graduate from each phase of training. That ensures standards and performance are immeasurably higher than the near sausage-machine pipeline historically apparent in some mass conscript armies. Training standards' utility depends entirely on what they demand and how well they are enforced.

As with all else, how well training is done is about leadership.

An essential but possibly unfashionable component of training is the use of shame, the humiliation or distress caused to an individual or collective failure to achieve the required performance standard or similar. This does not mean bullying or victimisation is anything other than abhorrent. Still, it does mean the graduated and increasing demand for the required level of performance should be held over the recruit's heads if they ever wish to progress or be deemed untrainable and discharged. The utility is that few recruits want to be one of the few that failed to meet the standard. This is completely different from the training for Commando or Special Forces, where most fail, so the element of shame is substantially less to non-existent.

While there is ample room to discuss the standards, how the standards are achieved must conform to the limitations of cost, time, resources, and wastage rates. Physical fitness programs cannot injure recruits so they are lost to training. As previously stated, basic training must produce a psychologically more robust soldier, substantially more calloused against discomfort and deprivation than the civilians they once were. That means training needs to be stressful. Standards should require mental and physical resilience to achieve. Soldiers who can march long distances at night in the rain and still be able to maintain high levels of personal discipline are of higher value than those who cannot.

Understanding how to build up that stress without causing psychological harm or breakdown requires some skill and experience, as well as allowing for time to recover and repair in one 24-hour period a week when in camp. Constant instructor presence, time pressure, appropriate denial of sleep and exposure to adverse weather conditions are all necessary, as is a credible threat of harm or injury if drills and skills are not rigorously adhered to. During live field firing, being in the wrong place at the wrong time could result in death or injury. If a recruit does not need to concentrate to prevent that, they simply don't need to concentrate. Training that does not demand attention to detail and high levels of concentration is nearly useless compared to that which does.

The Shoot, Move, Communicate and Sustain framework can usefully provide some examples. Let us now look at some simple test conditions.

For movement we want every infantry soldier to be able to carry 23kg (50 lbs) plus their weapon and required water 18km (11 miles) in three hours in temperatures not exceeding 20C/68F. This soldier must be able to do this on two consecutive days. With no equipment, they must also be able to climb an 8m rope using arms only twice in two minutes. If required, all infantrymen or men (or women) within an infantry unit should be able to prove they can perform these tests daily, with no special training or preparation. Adding land navigation using a map and compass also makes sense, albeit that would be a separate and additional test. Physical fitness standards for the infantry should be logically higher than those of other arms or services.

For shooting, every infantry soldier should be able to engage a 75 x 45cm steel target from the standing position with a weapon and optics appropriate to the task. On the word command, he needs to move forward one meter and hit the target with one or more rounds in 5 seconds. Kneeling at 200m in 8 seconds, prone at 300m in 10 seconds. This test is simple, cheap, quick to administer, and does not require complex range facilities with reactive targets and electronic scoring. Additionally, you may want to test the soldier's use of other weapon systems within the platoon, such as a light machine gun, pistol and grenade launcher, in a similarly simple way. There may be a demand to test marksmanship in NBC or CBRN kits and using night vision systems, but why test this when useful exposure and familiarity may suffice? Meeting a provably useful standard is good enough. Conditional to all the above is that the infantryman is tested in safe weapons handling, maintenance, and the law of armed conflict as when to shoot is as vital as how to shoot.

For communication the soldier needs to show they can assemble and power up the communications equipment, select the required channel and encryption settings and send and receive voice and data as appropriate.

Sustainment is a combination of combat medical training and living in the field. Medical testing is well-trodden ground, but living in the field can also be subject to simple standards. For example, you need someone who can wake up in the field, feed, wash, clean their weapon, and be ready to move in less than 30 minutes. If they cannot do that, there will be many other things they cannot do, or they will be substantially less useful than those who can. Teaching platoon or section tactics to soldiers with good personal skills, administration and physical fitness is fairly simple and easy to do.

Everyone can argue about the standards above, but in broad terms, many other benefits follow if you have soldiers who can do those things. As previously stated, standards will vary between arms and services, but there must be standards and tests that are pass or fail. The tests should be simple and very cheap to administer. The more pressing question is, if you assume your job is to prepare soldiers for warfare in the 21st century, why would you not want to set demanding standards to exclude those incapable of gaining and maintaining them?

Unit Level Training

The high training costs start to be incurred in unit-level collective training, particularly unit-level field firing so gaining maximum benefit from limited field training is essential. Time and cost are serious constraints on unit-level training. Regarding a volunteer army, nights "out of bed" or days in the field need to be limited, or soldiers will simply leave the army. Even in a conscript army, cost alone means time dedicated to field training has to be limited compared to that of being on operations or just as a percentage of mandatory service time. The generally accepted number of nights out of bed for the British Army is, or was, about 90 days. The Israelis don't have a similar concept, but annual reserve service after initial conscription was traditionally limited to 30-45 days a year unless deployed on operations. The figure to be discussed in 90 days is based on the idea that 90 days of field training should deliver any unit ready for operations at immediate notice. Not all training is field training. In camp, training is just as essential for all the minor skills, drills and cadres a unit needs to run. Cadres and courses almost always have a field training component, but we should not overcomplicate the discussion for now.

The focus of unit training should be graduated from platoon and troop, through sub-unit to eventual unit field training where the ability to perform to the required standard can be tested. Testing collective training is a contentious subject, usually for reasons that should not exist, but hierarchies and bureaucracy often do. There is a tendency in modern thinking to separate how something gets done from its outcome, which is as much a negative tendency as focussing on the process and not the result. In almost all military activity, how something is done in terms of performance is indivisible from a better outcome than if something was done badly for an equivalent outcome. For example, how a unit gets itself across an obstacle quickly and in good order can be done within the daylight and with all callsigns using their communications. That is not the same as crossing the obstacle at night in bad weather, with no lights and radio silence.

The point here is that any unit-level activity that needs to be tested should be tested against set levels of performance, not processes and outcomes. Constraining time, both in terms of planning and execution, is the two things most likely to force the required performance levels. Box-ticking processes and procedures must be avoided, particularly regarding the orders process or an idealised sequence of events often trampled by reality. For example, you may need to conduct a river crossing without input from the Engineer's recce detachment because time simply did not allow it. Units can only be as good as their platoons/troops and sub-units. Understanding what good looks like is, therefore, critical.

It is reasonable to assume that, except for medical and maintenance units, everyone deployed in an assembly area or hide should never fall below 30 minutes' notice to move, as the unit should be ready to move within 30 minutes of being ordered. If that unit has been stationary for 72 hours in freezing rain, then only the highest levels of discipline and maintenance will mean all vehicles will start when required, and all equipment will be packed away and stowed in time. Unit-level field training is where these skills are developed, and it is from extremely simple but demanding levels of performance that wider skill sets are developed.

Ensuring a unit can remain undetected within a hide or assembly area was something constantly drilled in the Cold War and is still very relevant today, so constantly checking a unit's level of signature by using divisional or formation sensors will rapidly develop the required skill set for units to avoid detection.

So what about all the armoured vehicles shooting, moving, and communicating?

Cost is the limiting factor for live firing, whether on the range or in the field. The maths is instructive. Assume you have 48 x 155mm guns. Assume your planned warfighting daily rate per gun is 350 rounds, and you want to hold 30 days of warfighting stocks. This means you need to have 504,000 complete rounds on hand at any time.

Assuming a shelf life of 10 years, this means that 48 guns will have 50,400 rounds a year to train with. This equates to 1,050 rounds per gun.

At current market prices for complete rounds, as in shell, bag charges and fuses, this equates to about $6,000 per round.[1] Each gun will consume $6,300,000 of ammunition. Thus, what training value do you want to get from 1,000 rounds per gun per year or just over $300 million in 155mm ammunition annually? Remember that if you want to fire at war usage rates to train your gun crews and artillery logisticians, then 1,000 rounds is just three days of firing. Limit that to 100 rounds daily, and you get ten days of firing comprising 16 full-rate fire missions. The important question is not just how many rounds a year a gun crew have to fire to be competent but how many live fire missions a forward observer and fire direction centre control to be equally as competent.

Direct-fire ammunition costs for cannons and tank guns can be substituted for tracer projectile target practice rounds, which are substantially cheaper than operational rounds, such as 35mm x 228, quoted at $600-1,000 per round.

Pedants can debate the various sources of ammunition costs. Still, the simple fact is that a live-firing battle group attack with armour, infantry, mortars, and close-support artillery, as once done by the British Army on the Canadian prairies or by the US Army at the National Training Centre, consumes vast sums of money in terms of ammunition.

In addition to the ammunition costs, the size of training areas and their attendant safety ranges is a significant constraint. In very real terms, unit-level live firing may be prohibitively expensive for most modern armies. Suppose the unit focuses on high levels of crew skills, which can be tested and confirmed via annual training. In that case, the requirement to bring these all together in combined arms or formation-level exercises might be better done using laser and GPS-based simulated engagement equipment, especially concerning force-on-force training. Training cannot be agnostic of cost or be based on simplistic demands for things no longer affordable or possible given current political conditions.[2] Much of the answer here lies in simulation, which must be correctly applied, like any tool. Can simulation reduce the amount of live firing required for vehicle gunners? Yes, it can, but the gunner will still need to fire some live rounds to confirm the outcome of the training. Can simulation reduce the number of days a unit needs to spend in the field? No, it cannot because simulation cannot teach all the awareness, skills, and determination it takes to live in the field and operate and maintain vehicles and equipment any more than simulation can benefit a football player. Most armies recognise this to varying degrees, but simulation advocates promote it as a tool to save costs, not to gain more benefits from limited training time and resources. My experience has been that those who most strongly advocate for simulation from an industry perspective seem to have significant gaps in knowledge and experience, leading to fundamental misemployment.

The required ratio of simulation to live training is the one that gets the required standard for the least cost. It is not the one that leads to a reduction in field training or an overall reduction in cost. It should be obvious that simulation makes possible training impossible under live conditions, as in force-on-force training. This cannot be discounted as being immensely valuable for no other reason than simulation will more probably demonstrate warfare as it is, not as one would wish it to be. Given the correct physics modelling and approximation to reality, simulation can and has exposed the real flaws in the conduct of warfare as many have come to understand it.

Simulated battles and engagements also allow for research otherwise impossible to conduct and usefully apply human comprehension and understanding of concepts and doctrine to be applied in some practical form. Many observations used in the writing of this book were derived from simulation and wargaming, not to see into the future but to understand better and test current ideas and teaching, much of which seems less than valid given the most likely circumstances.

Wargaming

Allied to simulation, Wargaming can be an extremely effective training and force development tool. Recent years have seen a huge interest in wargaming, with many universities and academic institutions choosing wargaming as a "method of enquiry." Professional Wargaming used in the military should be very distinct from hobby-type or academic type wargames. The IDF defines military professional wargames as "a bilateral (or multilateral) simulation of a military activity that represents real or hypothetical situations. A war game is designed to examine operational ideas, assimilate plans and analyze concepts and systems according to defined rules."[3]

Additionally, the following characteristics should characterise professional military wargaming:

- Use the information that drives real life military operations decision-making
- Use real maps, of large areas, of real terrain

- Use real planning, staff, data, orders and procedures
- Use real people, using their training
- Use detailed tracking of logistics and casualties
- Use multiple games and multiple teams of players to verify insights.

This differs from a board game with dice, counters, and a hex map. The wargame usually has no professional relevance if commanders do not make decisions based on real-world data.

Wargaming is a powerful tool, so it must be treated cautiously. Hobby wargames have tangential military and social value, so they are not entirely useless in providing insights that may have a professional application, but competence in hobby wargaming may not demonstrate competence in their real job. Professional Wargaming should demonstrate real-world competence. Professional wargames require professional knowledge and skill, in the same way, yacht racing requires extremely skilled seamanship and boat handling long before any racing is possible. If you cannot plan and write orders for a real HQ, a professional wargame should not forgive or allow that handicap regarding professional military training and force development.

We will not cover how such games are executed or conducted here. Still, it suffices to say that high levels of training and force development insights can be accomplished with little more than the appropriate staff, commanders, cartography, data, military knowledge and judgement. Very little in the way of technology or networks is needed.

The critical point to understand is that wargaming has historically been as poorly applied as it has been used to create success. The US Navy's use of wargaming to thoroughly understand how to defeat the Japanese in WW2, developed in the 1920s and 30s, is a testament to a rigorous system thoroughly applied as is the Royal Navy's Western Approaches Tactical Unit to defeat the U-boat threat to convoys. The British Army effectively employed a system known as the Divisional Wargame for the force development of the British Army of the Rhine. Conversely, much literature and mythology cover the variable conduct and execution of the US Joint Forces Command Millennium Challenge in 2002.

For a wargame finding or insight to be valid, it must replicate across a series of games and evolutions and, ideally, across varying forms of simulations and methodologies. Like hard science, if you cannot replicate results, something is wrong with the wargaming methodology or the overall approach to the research or training.

Insights gained from one execution of one wargame are inherently unsafe and lack rigour. Multiple iterations using multiple players are needed. Given a realistically modelled enemy force with a competent enemy commander, wargaming can be used to test and command a staff's ability to produce competent plans and test their execution.

This has major training benefits, particularly in forcing commanders and staff to write orders, especially when given incomplete or conflicting information. Wargaming should force the requirement to generate executable orders quickly and can generally indicate the level of competence a commander possesses.

Wargaming's utility is entirely predicated on the right people applying the right methods for the right reason. Any variation on those three requirements will produce dangerous and misleading results.

Divisional and Formation Level Training

Large-scale exercises outside military training areas were as common during the Cold War in West Germany as in most nations in the early 20th century. They were vastly expensive but did provide a wealth of experience and insights to the commanders concerned. Today, these would be almost impossible to conduct in the same way. Still, a lower-impact variation of such exercises could be conducted using civilian vehicles and limited to command staff supported by simulation and virtual battlefield-type technology and limited numbers of troops on the ground. The aim would be to provide staff and commanders with the realities and frictions of applying their orders to real-world problems against a sentient enemy on real terrain with real consequences. The application of such an idea might be units with each platoon, troop or detachment represented by one civilian vehicle with 4-5 officers and NCOs. Thus, a unit would have 10-15 civilian vehicles, which would simulate the execution of orders within some type of framework to allow umpires to decide outcomes and events using a combination of Wargame and Staff ride processes. The decision on the value of this type of activity would have to be viewed as the only one possible within budgetary constraints.

Training Value

Training can be good and bad, but what differentiates one is the relevance of what soldiers and commanders know or can do. An infantry platoon that can conduct reconnaissance patrols in all weather and at all times of day for sustained periods probably has little else to learn. A Battlegroup HQ that can routinely produce a full set of orders in one hour is probably as good as it is ever going to be. It has long been espoused in the British Army that even Special Forces training is merely "basics done well." The entire training ethos of good armies should seek to do the same. In many ways, the traditional Israeli approach stands testament to that as with limited time and budget, the only option is to do basics well. There is little merit in trying to be clever and complicated. Basics do not mean simplistic or easy. Humans tend to be fascinated by uncertainty, as in the endless possibilities of chance. Brief any tactical drill and concept; usually, someone in the audience will ask, "But what about if the enemy does X or Y?" X or Y is often highly unlikely, but the sense of uncertainty requires some comfort because no one wants to hear that "both parachutes might fail." However, accounting for it slows the action needed because it creates more complex decision-making.

The training that produces real value is the most utility and relevance for the time and money spent achieving it. That usually means identifying those few overarching individual, sub-unit, or unit-level skills that are constantly performed to a high standard and have benefits not achievable by other methods.

Conclusion

Any attempt to save money in training without being able to quantify or measure the compromise being made is bound the create an army less ready for war. The budget better spent to eke out every last ounce of value from an adequate training budget will save lives in combat and result in mission success.

Training should be hard and emphasise the cost of failure and why not being good at your job is mostly unacceptable for anyone to tolerate. Training should constantly test all concerned and lead to the removal of those unable to meet the standard, be that a recruit or a divisional commander. The reward for being well trained can only be discovered in war, thus a reward no one should seek unless forced.

Endnotes

1 We can debate exact costs, but this was a reasonable figure at the time of writing to include the shell, bag charges and fuse.
2 In 1984, the UK spent 5.5 percent of GDP on defence.
3 IDF Dictionary of Military Terms 1998, translated and transmitted by Israeli Operational Analyst via Email. 2018

5

INFANTRY

The next few chapters will examine the various arms and services that could exist today in the Divisional model discussed in Chapter 1. Many opinions will differ based on the Army the reader is familiar with and their cultural touchpoints, such as the British Army's "professional NCO" model, which differs greatly from the Israeli or German armies regardless of merit. This book is not a debate on culture, so while critical to the army's concern, it bears little on the following discussion.

Infantry is the lowest level of signature for a manned system on the battlefield. Infantry is the ultimate precision weapon, able to apply lethal force precisely, proportionately, and discriminately even within rooms containing the enemy and civilians if required. In certain terrains and environments, the infantry is the best form of reconnaissance. Yes, miniature UAS may have a role. Still, when they can't get past a closed door, thick foliage, or even bead curtain. They probably can't fly under the canopy in the jungle or forest and don't work well in the desert with a constant 10-20 knot wind.

In sharp contrast, Rifleman Snotgoblin will mostly do his job regardless of those conditions. The well-trained infantryman is the most flexible entity in most armies' inventories. In broad terms, the modern infantryman should be the primary means of reconnaissance on which all else relies. Give the infantry the sensors and communications to access every fire and effects support available. Technically and conceptually, this is the current state of the art.

So, what does the modern infantry do?

Ask the internet, or any collection of retired officers and SNCOs, what the role of the infantry is, and the answer will be some variation of the idea that the role of the infantry is to close with kill the enemy. The acme of my infantry training was that you needed to be able to jump into the enemy trench and bayonet the enemy to death, even assuming he was mortally wounded from the grenade you had posted into this trench before the assault. Has warfare changed to the degree where this is no longer true or where the capacity to perform such actions is no longer needed? It would be a foolish man who suggests this was the case, but historically, it was very rare that this ever occurred. Yes, infantry should be able to engage in dismounted close combat when and if required, and the training burden and demand should not be lightened in this regard, but that is not to say better ideas and better training should not seek to reduce both the likelihood and risks of close combat.

What follows supposes we train, equip, and organise infantry around the overarching concept that they need to access any point on the battlefield to employ weapons, communications, and sensors and have a line of sight to any other point on the battlefield. In that case, we might describe the infantry's role as "enabling the defeat of the enemy by gaining such proximity as to break their will to resist." Then, this might be the basis of progress. That may still mean "close with and destroy", but the infantry's raison d'etre should not be to engage in hand-to-hand combat with the enemy on his defensive position, nor should that be an objective sought by training. In terms of direct fire, the closer the infantry get to the enemy, the more effective their weapons become. If you can close the distance to the enemy and the enemy cannot prevent that, this will create defeat.

The Monash Infantry Company

Much has been written about infantry organisation, including by this author, but it would serve no purpose to restate the arguments and debates here. The arguments as to whether you want three platoons of 32 or four platoons of 24 and how those various platoons should be organised and equipped are entertaining and raise the occasional good insight. Still, they are ultimately unresolvable and most funded investigations done by armies are usually predicated on trying to prove or find evidence for why manpower costs could be cut or why they should not. Overwhelmingly, when I have sought to address this in discussion with other subject matter experts, three apparent conditions overturn all the other details.

First, the effectiveness of an infantry sub-unit, platoon, and section is defined by well-trained and robustly selected junior officers and NCOs above all else. Training and leadership trample all other factors.

Secondly, operational conditions, mission type, casualties and manning will demand the sub-unit, platoon and section be able to function in multiple forms of organisation. So, three platoons of 32 may break down into six multiples of 16 comprising 4 x 4-man fire teams and then regenerate as four platoons of 24. Organisations must conform to what is best suited to the mission or operation. Table of Equipment and Organisation Charts do not survive into reality. They exist because there must be a pay, administration and procurement baseline.

Thirdly, while organisation and manning are vital for budget and administration, casualties and chaos will strip away that certainty at every level. You may have trained to fight with a 24 or 32-man platoon, but on the day, you will have 16 men, of which only two will be trained and experienced section commanders.

The baseline Monash platoon consists of 24 men, one officer, and twenty-three other ranks. Three such platoons exist in a Rifle Company, for a total of 72, comprising four officers and 68 other ranks, exclusive of the Company commander and HQ. There are four rifle companies in the unit.

Each platoon has three NCOs, each trained and selected from a course of training that enables them to train all the individual soldier skills, plus platoon weapons and command a section of 8 men. Each platoon has three 8-man sections: a command section and two rifle sections. All soldiers in the platoon are equipped with 5.56mm modular individual weapons. The Officers and NCOs all carry 0.25-5-watt hand-held/body-worn IP-based 30-512mhz radios capable of ad-hoc IP networking. If the budget allows, you can issue every soldier with a personal role radio, but barring cost, this item seems to have little impact on overall training and organisation.

The platoon commander leads the command section and breaks it into two four-man teams. The Platoon commander's team comprises the platoon commander, a platoon medic, and two soldiers who can assist either the medic or the commander. They might operate a platoon unmanned Air or Ground Vehicles (UAS or UGV). The other team comprises a Sniper pair, a signaller and a specialist NCO observer who locates and generates target data for supporting fires. That ability is inherent to the training of all four soldiers and not the sole responsibility of the NCO. That location data and target description are passed back to the attached Artillery Forward observation officer within the Coy HQ.

The other two rifle sections are composed of two fireteams, one equipped with a belt-fed 7.62mm light machine gun, which can be swapped out of an 84mm recoilless rocket launcher if required. The other team carries individual weapons except for the section commander, who will optionally equip his IW with a 40mm grenade launcher.

We will not delve into platoon weapons more because no army has ever lost a war because they had the wrong platoon weapons. The important point is that the communications equipment and weapon sights are probably far more important than the make or type of light machine gun or sniper rifle. A platoon with digital communication and a quantity of thermal weapons sights will outperform any platoon not similarly equipped. The infantry equipped with IP-based software-defined radios will access supporting fires with greater ease, rapidity and certainty than one which does not.

The company HQ comprises one officer and a senior NCO. The officer is primarily concerned with fighting the rifle platoons, while the SNCO looks after the Company support echelon concerned with sustainment (Company Sergeant Major). No "2IC" (second in command) exists at the Sub-unit level. The Company commander has a team comprising one signaller, an attached Artillery officer, and his signaller. The Company Sergeant Major (CSM) has a team comprising another Senior NCO, four cooks, four general duties men and two signallers. These men are trained infantrymen, except for the cooks, who are attached specialists.

Load Carrying

Before examining the unit structure, we must consider the carried load as load carrying is another subject dominating the low-level infantry debates. Many extant studies cite either historical experience or trials and experimentation data, most of which have been wasted by failing to address the simple truth that only badly commanded and badly trained soldiers get overloaded. This bears repeating and emphasis. If your soldiers are overloaded, there has been a failure of command and training. It is that simple. No load-carrying study I have ever seen makes that point. There may be good reasons to explain why the soldiers are overloaded, and orders from higher command may have mandated what needed to be carried and why. Still, the higher command is at fault, as are the junior commanders who failed to speak out. Load carrying is not and never has been a weight problem. It is and always has been an ideas and leadership problem.

None of this means that how an infantry sub-unit is organised or what it is equipped with is unimportant. The opposite is true, but it does mean that debates and discussions must be built around certain core fundamental ideas on which all concerned agree.

Let us start with carried weight because no man can fight with empty hands. What a man can carry in the dismounted fight pretty much defines the building block around which all else can follow.

Weight is finite, so the insights and improvements lie in understanding what is not required and not carrying it. The traditional approach to this idea has provided commanders with a conceptual framework of carried items, usually arranged around three tiers described as "assault order" or patrol and marching order. This does not, and never has been, a substitute for the experience and judgement required by commanders.

Hence, the utility of such frameworks is essentially a last resort when experience and judgment are absent. The tyranny of carried load is the lack of information on which commanders' decisions are made; thus, most carried weight is guarding against uncertainty regarding mission duration and type. It is possible for a short-term operation planned for a couple of hours to extend to 48 hours or more. Yet infantry does not and never should operate in isolation from support. The load-carrying debate can be solved by re-supply and distribution at the sub-unit level, but that must be confirmed by strict enforcement of load restrictions. Load carrying is inherently connected to unit logistics.

Soldiers must work within weight limits so that time, risk and resources are well used on resupply. 8 x 24-hour ration packs and 10 litres of water weigh 16kg, easily carried by one man, compared to the approximately 50kg of 5.56mm, 40mm grenades and 7.62mm link needed to resupply a section. There is also the fact that if you cannot move re-supply forward, you almost certainly cannot move casualties back. Any army can reduce carried weight if it allocates manpower and equipment to platoon resupply at the sub-unit level and commensurately the same at the unit level.

Carried load means making hard choices and running some risk, but war is inherently risky. The issue of risk leads to the issue of personal protective equipment such as body armour and helmets.

The statement that if wearing body armour and or helmet means you are less likely to accomplish your mission, then it should not be worn is a useful insight, but that in no way detracts from the fact that body armour saves lives, even to the degree where in previous conflicts survival would have been impossible. Combat deaths can significantly impact the political will to sustain a force on operations. If body armour weighed nothing, then there would be no discussion, but hard body armour cuts into carried weight significantly. It can also cause significant heat exhaustion, as the UK experience in Afghanistan and Iraq showed. For example, the current

UK body armour and helmet have a combined mass of 8.25kg, or just over 18 lbs. What compromises body armour varies considerably, be that low coverage hard armour against rifle rounds or high coverage soft armour against light artillery fragmentation or the various combinations of both, is something that operational analysis can inform and should do exclusive of political pressure. Soldiers willingly carry weight for something they feel or know has proven value. They will reject that which does not, if allowed to do so.

Ultimately, the decisions regarding carried weight are about leadership and training. Well-trained soldiers with good leaders will not be overburdened, except in extremis.

Infantry Support Weapons

The focus of the discussion here will be limited to non-vehicle mounted systems, so employed dismounted. Paramount in this area are the anti-armour weapons and mortars as the two support weapon types that should attract most investment. The potential efficacy of effective anti-armour-guided weapons held by an infantry unit should not be understated. Current western systems such as Javelin and Spike are easily man-portable and have effective engagement ranges of 4km for the Javelin with the Lightweight Control Launch Unit (CLU) and 5.5km for the Spike LR2 variant, which uses fibre optic guidance, so it is a non-line of sight capable (NLOS). The French Akeron-MP is like Spike in that capability. The important insight about NLOS is that guidance does not require the detailed sighting normally associated with direct fire systems, which means fewer systems can cover the same frontage as direct fire systems.

The Monash infantry unit would aim to be equipped with 12 Akeron-MP or Spike-type systems, each supported by three men. This would be one officer and 39 other men, comprising the Officer and SNCO, plus two signallers in the command section. Then, there would be six sections of six men, each with two posts commanded by a JNCO. Assuming a 5km maximum range, this is enough for a 24km frontage as each pair covers 8km.

Direct fire systems sighted in pairs might require as many as 36 systems/posts. That not only has significant cost and manpower implications but also requires a lot of redundancy in where to distribute and allocate the spare rounds and the detailed sighting of each pair of posts.

Anti-armour-guided weapons should not be viewed as weapons to be used only against Main Battle Tanks but against any similar high-value target or weapons platform.

Such weapons also provide a precision fire capability against legitimate targets in close proximity to civilians, which should not be underestimated in terms of utility.

Modern mortar sighting and fire control systems mean individual tubes only require GPS to provide an initial fix as to location. They can then fire accurately and rapidly on any target within range. This means mortars can be dispersed against counterfire and are substantially harder to detect and easier to conceal. Airburst fusing makes all calibres effective anti-personnel weapons. The other flow-down effect is that modern fire control systems make mortars far quicker and simpler to react to targeting data provided by the observers.

In terms of effectiveness, the 120mm system outstrips all the others in terms of ammunition performance because guided rounds using either semi-active laser or GPS/INS are now mature and proven technologies which have been combat-proven.

Similar systems do exist for 81mm but are far less common. The problem is that while effective dismounted from vehicles 120mm requires a vehicle of some sort to move and sustain. Given the choice of three systems, it would be easy to assume that these somehow correspond to a light, medium and heavy force structure, but this is not wholly correct. The last decade and a half have seen 120mm mortar systems be mounted on increasingly lighter and lighter vehicles either using very high-performance recoil attenuation of vehicle-emplaced baseplates, which means 4x4 Light utility vehicles can rapidly deploy and displace 120mm mortars as well as carry a useful first line ammunition load. It is certainly viable to support light force with a 120mm mortar. In jungles or mountains, which are impassable to vehicles, 60mm is the only option. It has the added benefit of having "commando mortar" or handheld variants, making the carried load far from the impediment a tripod and baseplate mounted system might be.

Which mortar system best suits the task will be a function of cost, weight, and complexity. Objective metrics of terminal performance on various target sets may help inform that decision. Still, ultimately, costs in training, manpower, and procurement will drive the decision, as will the costs associated with supporting the system's weight across the unit of formation. A 60mm mortar may not impact a unit's organisation and vehicle allocation, but a 120mm mortar certainly will.

The Monash 120mm mortar platoon is organised similarly to the Anti-tank platoon, as in six tubes on but crewed by three men plus a driver and commander when the vehicle is mounted. Wargaming for this work strongly indicated the need for all vehicle-mounted 120mm systems to have dismounted capability. This covered many possibilities, including losing the vehicle, positioning the mortar within field fortifications for reduced signature, and enhanced survivability from counter-fire.

What can be stated with reasonable certainty is that infantry subunits supported by NLOS-capable anti-armour guided weapons and mortars mainly focussed on creating anti-personnel casualties via high explosive fragmentation have very little to feel is lacking from their inventory.

In this example, significant benefits can be gained from employing unmanned systems, such as UAS and UGVs, which form the only other unit support grouping. The IDF issued all infantry units with UAS sections in 2008. UAS has been a key component of land operations since the 1980s, so despite all the media excitement, the Russo-Ukraine War from 2022-24 has not seen the revolution claimed because, as of February 2022, neither side possessed nor had much experience of operating UAS at a force level. Thus, the UAS systems present, such as the Bayrakat TB-2, tended to have a disproportionate effect. In sharp contrast, as of 2013, the British Army in Afghanistan routinely operated the 16cm long, 18g, hand-sized Black Hornet, capable of a 20 mins flight out 1,000m. The British Army understood the value of small tactical UAS nearly a decade before Ukraine or Russia.

The requirement for a unit-level platoon to operate some type of UAS dedicated to electro-optical target finding has existed for over a decade. Unit-level UAS needs are austere regarding capability. If the primary role is to support the mortars and guided weapons platoons, it only needs to detect and identify targets out of the 10-12km slant range. The systems need to operate in a highly contested EW and GPS-denied environment and be man-packable by two men or maybe 4 men but limited to one light vehicle, so the same level of debt as a 120mm mortar. Short endurance civilian specification "quadcopter" type systems might be issued on an as-demand basis and have no real impact on the manning or platform count of the unit. Depending on cost, they might even be viewed as disposable as a munition to be consumed rather than a piece of equipment.

Most capable military UAS systems are substantially more capable than the unit-level requirement might demand. This means specialist formation-level UAS detachments may make more sense from a balance of investment perspective. This will be pursued in detail in another chapter.

The real need for unit-level UAS detachments may not be target detection but logistic distribution. As of August 2024 electrically powered quad-copter type UAS can lift 180kg of stores 20-30km forward, land, be unloaded and return to be recharged, reloaded and re-tasked. The growth in the supposed market for "people carrying drones" and drone air taxi services means that the technology possibilities for UAS to become a viable method of resupply at weights of 2-300kg are now looking more viable. This does present the potential for UAS to be used for casualty evacuation.

The UGV requirement is likewise more problematic than popular perception portrays, but UGVs have significant potential. UGVs have two proven and mature capabilities: logistic distribution and re-supply, which combine with casualty evacuation.

The less proven tasks are reconnaissance and weapons carriage.

Regarding logistics, that indicates a need for some small and light vehicles to operate forward of the "trailhead" or "forest edge" to sustain dismounted manoeuvre and casualty evacuation. Manned vehicles capable of this type of task exist and have done since the end of WW2. Prominent among them is the M274 ½ ton 4x4, which weighed 369kg but could carry 450kg in a platform 1.27m wide, 3m long, a 170km range based on no more than 30 litres of petrol. More than 11,000 were produced and used extensively in the Vietnam War. In the mid-1960s, the Bundeswehr procured a similar, more capable, and complex vehicle in the shape of Faun Kraka 640 with a load capacity of 750kg.

Modern manned platforms such as the Multipower Hippo HAWC (hybrid amphibious wheeled carrier) and the Supacat ATMP (all-terrain mobility platform) correlate more strongly with the proven concept of the M274. This poses the question of why a robotic version of the same platform is required. Autonomy is neither required nor desirable, so all these vehicles require drivers, whether on the vehicle or dismounted with a control terminal. UGVs need to be under human control and protection to prevent enemy or malign civilian interference and ensure the most effective execution of the task.

The obvious debate is that while a low signature, all terrain logistic support system for dismounted infantry is highly desirable, why must it be robotic with the attendant and added cost weight and complexity?

Taking the man off the platform may be desirable for the reconnaissance or weapons-carrying role, but removing the man does not mean he doesn't have to maintain some proximity to the platform. The case for a 3-400kg UGV in either the reconnaissance or weapons carrying role would be move sensors and weapons beyond that could be effectively utilised by a dismounted man. Regarding reconnaissance, is a UGV more or less likely to detect the enemy than a well-equipped infantryman? Can the UGV access the same terrain and move through structures? Can it wade a 1m deep river or move through thick undergrowth? The same limitations apply to any vehicle carrying direct-fire weapons. The Monash weapons-carrying UGV is equipped with a remote weapon station, usually found on armoured or light vehicles, and uses a .50/12.7mm HMG or 40mm automatic grenade launcher. Manoeuvring these weapons in support of dismounted infantry engagements has obvious utility compared to a manned light armoured or unarmoured vehicle mounting the same weapons because of the reduced signature and risk of exposure. 40mm AGLs have an armour-defeating performance far more than .50 and can high-angle indirect fire to distances greater than 2,000m. Some ammunition natures can perforate up to 50-60mm or rolled homogenous armour. Almost all the battlefield conditions imagined for when it makes sense to support the infantry with either Land Rover or HMMWV mounting weapons make far more sense to use a UGV, albiet that the UGV may need to be mounted on a trailer towed by a Land Rover or HMMWV to arrive at the point of application.

A Monash unit would have a UGV platoon capable of equipping and sustaining 4 UGVs per Company. Two UGVs would be weapons equipped, and two would be resupply and medical evacuation. The weapons would be 40mm GMG and 12.7mm equipped. Manpower to control each UGV would be sourced from Coy HQs and Platoons as the mission required. Every Coy-level UGVs detachment would be a five-tonne truck carrying the UGVs. Each truck would carry 4 UGVs, a driver, and two UGV maintainers specialists. However, these UGVs are not autonomous and are being used to move loads that would otherwise be impossible to man-pack at an effective speed and distance. They will always be under human control with the human nearby.

So what about the Sniper and Reconnaissance Platoons? The simple answer is that they can exist if required, but within a constrained force structure model, there may simply not be the room, so a hard and possibly uncomfortable set of questions gets raised.

All of these are legitimate but herein may lie the insight. The Sniper and reconnaissance capabilities exist within the platoons, but not as separate entities. Hard constraints mean hard choices and working towards solutions and methods within them.

Unit Support

Modern infantry will need other services and capabilities to function effectively on the modern battlefield, which has little to do with weapons, sensors, and communications. Something as basic as feeding and catering needs to be considered. Combat rations are expensive and not ideal in dietary terms for prolonged use. Rations held in the unit or sub-unit echelon should be retained only when actual combat operations are ongoing and centralised feeding is impossible. Feeding is also critical to morale and overall health, so it cannot be discounted as something less important regarding a capability driver than medical support. Both are equally critical. Armies have been conducting centralise feeding for centuries if not millennia, so there is little new here bar the problem of security and signature. Field kitchens need to be well hidden from detection and avoid thermal signatures, set routines, and needless concentrations of troops static in one place. A sub-unit catering detachment needs to generate 2-300 hot meals per 12-18 hour working day. Catering detachments need sleep and time to wash up, clean, conduct maintenance and prepare for the next feeding cycle. Despite the need to avoid routine, troops must set mealtimes around which commanders can plan work cycles and administration. Centralised feeding should not result in long lines of soldiers queuing in front of field kitchens. Unlike in the past, modern mass catering and fast-food packing does allow for hot meals to be delivered forward, rather than attempting to centralise troops to feed them. This is the most preferable method concerning signature and security, especially in defensive positions or assembly areas and hide sites. The same is true for troops engaged in long road moves where feeding can be combined with refuelling and maintenance. The only complication in this regard is holding the required fast-food container stocks, which must be supplied in addition to stores and provisions.

In the Monash unit, each company has a four-man catering detachment with a dedicated vehicle, a demountable cooking rig, and a stores trailer. Each catering debt will have additional feeding requirements, such as the Unit HQ and/or the Support weapons platoons.

As previously mentioned, medical support is another critical area which demands sound thought and reasoning. The 20 years of combat in Iraq and Afghanistan saw most NATO armies make groundbreaking strides in their understanding of front-line medical care. Still, it would be a mistake to assume that work in the counter-insurgency and security force assistance world has much bearing on the delivery of medical support in warfare dominated by a toxic combination of possible constant movement, contested air movement and the deleterious effects of artillery fire. The overall choices inherent to medical policy will be dealt with elsewhere, so here, the considerations are limited to what an infantry unit can do to stabilise casualties to allow for the removal from the point of wounding to their centralisation for evacuation to a medical facility such as a field hospital. At the unit level, this entire area wholly depends on training, with very little being a function of equipment costs. If you want to save lives, the answer is simply to train as many people as possible to treat trauma casualties and stabilise them. Having section, team, platoon, and company medics has obvious merit until they are not there or have become casualties themselves. It is not unreasonable to suggest that the training time and resources allocated to casualty care should match those allocated to individual weapons training. Logically, this creates the need for soldiers who can train medics rather than just specialist medics who are the product of courses external to the unit. Those medical trainers can then man what the British Army calls the Regimental Aid Post, Unit Casualty Collection Point, where casualties move out of the unit and into the medical support chain.

The last component of unit support is the Logistics platoon, which would consist of about 8-10 flatbed trucks able to hold the required amount of stores to at least replicate every ammunition type issued to the unit, as well as enough fuel for the entire unit to march 800km and then refuel.

Infantry Mobility

The Monash infantry unit described thus far has been vehicle agnostic. It could be equipped with a wide range of vehicles, from unarmoured trucks to MICVs. Little about that choice would substantially impact the overall organisation, but the recovery and repair platoon.

This is because the Monash infantry unit is intended to operate dismounted. It is not intended to manoeuvre with armour. The reason for this is both simplicity and cost.

Thus, the recent trend of convergence that created manoeuvre units comprising two infantry sub-units in MICVs and two MBTs sub-units has been rejected in this work. The rejection is based on cost, weight, and complexity, plus there may be an issue of limited utility.

There are three suitable vehicle types for the Monash Infantry units:

- Trucks
- Wheeled APC
- All Terrain APCs.

Trucks do not require much discussion. Using a variety of modern configurations, it is possible to seat 16 infantrymen in light scales with weapons on a 4-tonne truck and drive them up to 800km a day. It is also comparatively simple to armour that vehicle to at least STANAG 4569 level 1 or 2 while not incurring any great cost in complexity or maintenance. Nine or 18-tonne trucks will afford more equipment storage space and higher armour levels. These vehicles will have limited off-road mobility but are only intended to move between a base and the rear areas and hide sites short of the actual combat zone. This could also mean the trailhead in mountain, jungle, or forest regions. How useful that is will very much depend on the context. The USMC used this type of vehicle fitted with an armour kit as an infantry carrier and towed gun portee in Iraq from about 2005 onwards, designating the Mk23 MTVR with an armour protection kit. The British Army's AT-105 Saxon was essentially a truck chassis and engine with an armoured body and a truly awful vehicle.

The wheeled APC would use the current Patria 6x6 as an analogue baseline, the Arquus VAB Mk3 or the TpZ Fuchs 2. The need is for an armoured vehicle that can march 800km, is crewed by two, can carry eight and is amphibious.

The all-terrain APC would be the BVS-10 or STS Bronco. To move 800km, a transporter would best move this type of vehicle, but it can swim and move across the ground where no human can walk. In terms of manoeuvre options, this is a hard-to-beat capability.

Thus, a Monash infantry sub-unit would have ten vehicles, comprising three platoons of three and one for the company HQ. The role of the Monash infantry vehicle, whatever type, is to get the infantry soldier to the point of dismounting into close terrain where he can manoeuvre. This might typically be the edge of a wood, built-up area or the base of an escarpment of a similar feature.

The reader's question is probably why we are not discussing MICVs or IFVs, heavy-armoured cannon-armed vehicles with anti-tank-guided weapons. The reason is simply cost and weight complexity on the one hand and utility on the other. MICVs are designed to move with main battle tanks or so many suggest.

Almost all modern IFVs mount anti-tank-guided weapons in much the same way that the BMP-1 in the 1960s, but there is a critical difference in employment. While little documentary proof exists, there is a reasonable conjecture that the BMP-1 was intended as an APC which could mount and fire the AT-3 missile in defence. The early models of MCLOS AT-3 had a minimum engagement range of 500m. To cover this, the 2A28 73mm smoothbore gun launches a projectile of near identical performance to an RPG-7, with a co-axially mounted PKT machine gun. Some sources indicate that Soviet anti-tank guided weapons platoons in BTR or Airborne units were equipped with RPG-7s for the same reason.

Simply put, a convincing technological case suggests that the BMP-1 was never intended as an IFV. It was an APC with an anti-tank system designed to be used when static in defence. Neither weapon could fire on the move. An antipersonnel round for the 2A28 smoothbore gun on the BMP-1 did not appear until the aftermath of the 1973 October War. The BMP-2, by contrast, was a more IFV-like design but did not enter service until 1980, significantly after the IFV concept had progressed in the West. The BMP-2 is specifically designed to engage low-flying aircraft and helicopters with a 30mm cannon enabled by 1PZ-3 air defence sight for the commander. Post-1973 experience convinced the Soviet Army that a 30mm gave the IFV an anti-air capability but axiomatically made it more flexible than the anti-tank dedicated BMP-1.

Much IFV development occurred without significant combat experience against a competent enemy and it is entirely legitimate to challenge the need for IFVs on a cost, weight and complexity basis. The IDF, who fought the 1973 War, rejected the IFV concept and developed heavy-armoured APCs instead. That is not to suggest that infantry carrying armoured vehicles do not need weapons, but that does not axiomatically equate to an IFV as we have come to understand them. As of April 2024, the KF-41 IFV chosen by Australia tips the scales at 50 tonnes, and the allocated budget of $13.3 billion gives a unit cost of $29.5 million for 450 of them. The KF-41 has a visual signature that exceeds a Leopard 2 MBT in terms of overall height.

Everything that kills tanks kills IFVs and even APCs. Every land force needs to hold substantial battle casualty reserves, and $13 to $29 million is too much for something to sit in a hangar. If you don't care about cost, weight and complexity, you will purchase small numbers of things you cannot afford to lose.

The training aspect of APCs and trucks is that if you lose too many APCs, you can adapt to using trucks as a reversionary measure far more simply than had your tactical doctrine been committed to IFVs with cannons and ATGMs. Regarding weapons, the Monash APCs would be armed with 40mm GMGs, using high-explosive dual-purpose rounds with an anti-armour perforating capacity of >50mm, making them superior to some 30mm cannons ammunition natures at 2,200m.

Conclusion

Much of what has been written here will not find agreement or support. Simplicity is always suspect to minds that see themselves as creative and innovative, yet simplicity requires deep understanding and coherent ideas. Infantry is the core of all military capability. It needs to be cost-effective, light, and simple. It cannot be expensive, heavy, and complicated. The infantry will suffer the majority of casualties and almost always form the majority of any land force.

The idea that the military should simply buy the best is not a basis for sound ideas or logical discussion. Doing so means spending far more on land forces than most nations can afford, and far worse, it means subjugating deep understanding to the advertising budget of defence companies. Given the constraints inherent to the force development model, what is proposed is nothing radical or complicated.

6

CAVALRY

By virtue of 3,000 years of horses being a primary means of warfare, the term cavalry is loaded with cultural baggage no better demonstrated than by the British Army's Royal Tank Regiment Officers seeing themselves as a distinct type of professional far removed from those officers in "Cavalry Regiments" operating the same equipment in terms of Main Battle Tanks (MBT) in the same role.[1] In sharp contrast, the Israeli armour officers have no horse-riding tradition, and their army came into existence well after mechanisation was established as state of the art. The IDF's "tank arm" sees itself solely responsible for operating tanks. As we will see, this has both an advantage and a potential shortcoming.

In most Western armies, the legacy of horse-riding cavalry persists in useful and less useful concepts.

In terms of what follows the word, Cavalry will merely denote the requirement for a force mainly equipped and required to fight and operate mounted versus that of the infantry who do so dismounted. Thus, the term cavalry is not intended as a sarcastic social commentary but an entirely functional one. The primary role of the Cavalry is reconnaissance, which is traditionally to find the enemy and screen your forces against enemy reconnaissance by also defeating the enemy cavalry tasked with the same job, thus "counter reconnaissance".

In some circumstances, cavalry's superior mobility means it can rapidly exploit battlefield conditions, leveraging shock and surprise to create and maintain freedom of action. That last sentence should be as pleasing to a modern reconnaissance commander as it would a 1905 cavalry officer. Still, while self-affirming and exciting, the main task may remain reconnaissance and counter-reconnaissance. There must be some caution in assuming that reconnaissance just means finding the enemy. It just as usefully means confirming the absence of the enemy.

If the idea that contemporary operations have created a "sensor-soaked battlefield" then the idea of ground manned reconnaissance for which cavalry exists would seem superfluous, yet at no point in history since the advent of the aircraft or the freedom of action required in both air and the electromagnetic spectrum has there ever not been a need for highly skilled ground reconnaissance capability. Manned or unmanned airborne sensors have significant limitations regarding what they can detect and their endurance and proximity to a task while remaining undetected and survivable. There are also critical tasks and information, such as ground conditions or interactions with civilians, that they cannot accomplish. Even a UGV is less than ideal to inform a unit commander where he can or cannot cross a river or whether a patch of forest is a useable hide site. UAS have a very poor track record in detecting well-trained dug-in infantry or irregulars, as both Operation Anaconda in 2002 and the 2006 Second Lebanon War showed.[2]

Modern cavalry manoeuvres weapons, sensors, and communications that are far more capable and effective than those carried by dismounted force. These characteristics accurately describe a main battle tank, which should not be surprising. When the cavalry or reconnaissance regiments of the British Army first mechanised, they did so with "light tanks." The most iconic British Army reconnaissance vehicle of the Cold War, the Scorpion, later to become the Scimitar, was originally marketed by its manufacturer as a "light tank" in the British Defence Equipment Catalogue of 1973 and much later. Today, many IFV chassis can mount turrets containing large calibre tank main armaments, essentially making them tanks based on General Montgomery's enduring definition cited in Chapter Two. This is not a new phenomenon. In 1979, the Argentine Army mounted an L7 105mm equipped turret on a German Marder IFV chassis to the "Tanque *Argentino* Mediano" or TAM. To placate inter-service doctrinal infighting such vehicles often get described as "tank destroyers" or "infantry fire support vehicles" but are functionally tanks, lacking the levels of protection inherent to most designs.

In terms of equipment, this creates an interesting debate: assuming the same sensors and communications, does the Cavalry need an IFV-based unit to be equipped with vehicles that mount 30/35mm cannons, anti-armour guided weapons and carry 6-8 infantry for dismounted close reconnaissance? Or using the same chassis to mount 120mm guns and 40 stowed rounds with 90 percent automotive commonality between the two. Would a good Cavalry unit configuration be two sub-units of IFV-based vehicles and two sub-units of "tank-like" vehicles? If readers think they may have read something near identical at the beginning of this chapter, then they are not mistaken. There is nothing new here in terms of approach. The French Army's Project Scorpion project is based on a 6x6 APC and 6x6 armoured car using the same major automotive components in much the same way as the UK did in the 1950s with the FV600 family of Saladin and Saracen armoured car and APC.

AFVs equipped with fire control systems, stabilised turrets, medium calibre cannons, or 105-120mm main guns are not cheap, although less expensive than main battle tanks. They possess a comparatively high signature and attract significant training, logistics, and maintenance support costs. Arguably this description falls into the heavy type of heavy, medium, and light taxonomy. However, it is worth noting that currently the CV-90-based vehicles used as notional type equivalents for the IFV and MBT have a stated mass of between 28-32 tonnes. The Australian Boxer-based CRV with a 30mm turret is reportedly 38.5 tonnes for an 8x8 wheeled vehicle.

This seems to suggest a clear demand for light cavalry with both low signature and high mobility able to employ anti-armour weapons for counter reconnaissance. The vehicle type might be little more than a modified 4x4 as common to most special forces patrol vehicles, through to more specialist types such as the UK's Jackal and Coyote that equip the current light cavalry units, which are currently demonstrative of this approach.

Light armoured alternatives stretching from the French Army VBL and the Japanese Komatsu LAV to the German-tracked Wiesel-2 are all possible candidates which may be deemed suitable, as in a vehicle with some armour protection against short-range small arms fire and light fragmentation, but just as importantly small enough to quickly and easily disappear under a layer of thermal camouflage netting with 1-2 mins of halting or to conceal the vehicle against or within a building. If the vehicle can park inside a twenty-foot shipping container then the amount of place it can be concealed far outstrips those of a larger vehicle. Equipping whatever vehicle type is selected with communications, a 5km NLOS capable ATGM and the relevant sensors to detect enemy vehicles and personnel at the same ranges adequality fulfils most reconnaissance requirements. If these sensors can be mounted on telescoping masts, then this has clear advantages for detecting the enemy at a lower risk of being detected.

All reconnaissance vehicles should be able to obtain and transmit Category-One grids to call for fire. Long-range and highly dispersed operations could well mean an HF communications fit of some sort.

Counter reconnaissance demands the application of anti-armour weapons. Still, light vehicles mostly negate cannon armament, although some 20mm and 30mm chain gun types can and have been fitted to light vehicle chassis in remote weapons stations.

However, the M230LF has completed trials mounted on a Toyota Hilux. Direct fire weapons are useful for short-range fleeting engagements in urban or woodland, forest, and bushveld, where trees will restrict guided weapons and only partial views of targets. As previously described, 40mm GMG is probably sufficient for light anti-armour and anti-personnel employment. Giving Cavalry forces significant or potentially useful anti-armour capability does raise the potential for such forces to form an anti-armour reserve once the screening force has broken contact and withdrawn to refuel and rearm. This could be the case if the defending force action were being conducted by dismounted infantry with a blocking or delay mission. The issue to be addressed is the extent to which the force wants to conceive of a "Reconnaissance Anti-Tank" unit as something with a perhaps paradoxical or conflicting mission set, yet a reconnaissance unit with a strong anti-armour capability is exactly that.

The same debate or discussion that supposed infantry fire support may best be applied by weapons-carrying UGVs can be applied to reconnaissance. As suggested for the infantry, the vehicles would be equipped with trailers able to dismount UGVs and carry capable weapons or sensors. The UGV could move forward about 1,000m under secure, low-powered IP-based communication. Additionally, such vehicles could remain static and concealed, providing sensor feeds across the network. Forty such vehicles could probably screen as much as a 50km frontage. The same would apply to manned vehicles, but UGVs should have substantially smaller signatures. If remoted via the D10 landline, the vehicles can operate with zero RF emissions.

UGVs potentially offer a bright new, even less arduous, operational environment but it must be emphasised that this offers no manpower or cost savings. A UGV reconnaissance capability still needs the vehicle crewed by 3-8 men, and a trailer to position and support it into action. The UGV merely offers a risk reduction methodology, which can reduce the need for direct protection, thus saving cost, weight, and attendant complexity. Everything the UGV can do un-manned requires the redundancy of a manned reversionary capability.

Cavalry units will still need to generate dismounted capability not unlike infantry, albeit as a far lower overall percentage of unit manpower.

Still, their dismounted forces should be strictly limited to reconnaissance tasks and not combat missions. That does not prevent them from using supporting fires as and when required. It should be noted that the discussion thus far explores the ideals and the reality that a resource-constrained model will have to account for. Having laid out the debate broadly, we must move to where real problems are tested against actionable solutions to get real insights.

The Monash Cavalry Unit

As with horse cavalry, the Monash construct defaults to heavy and light models within a doctrinal construct that defines cavalry as units trained, equipped and organised to operate mounted in vehicles. Despite what was discussed previously, the Monash Cavalry does not use UGVs or have a dismounted component because those elements are inherent to the infantry designed to operate in close terrain. Dismounted reconnaissance is an infantry task. The mounted task is intended to exploit open terrain unsuitable for dismounted manoeuvre. The Monash Cavalry's employment can be simplified to exploiting open terrain for manoeuvre and thus to counter the enemy in open terrain and prevent their manoeuvre.

The Monash Cavalry unit is organised into four sub-units of 10 vehicles. These can be organised into five pairs, with four reconnaissance pairs and one command pair, or three platoons of three with a commander. It doesn't matter which, as long as the soldiers are trained to do both or any variation forced upon them by losses. Each sub-unit is supported by two vehicles, holding echelon stores and a catering detachment. Each sub-unit has one Intercommunication light vehicle for administrative tasks.

The Unit HQ is configured like the infantry but has two additional fighting vehicles for the commander and his 2IC.

The next grouping is a Low-Level Air Defence Group of eight fighting vehicles and eight re-supply vehicles. While the Division has its air defence unit, this group is entirely dedicated to defending the Cavalry unit from air threats such as attack helicopters and loitering munitions.

The Medical, Recovery and Repair, and Logistics platoons are broadly similar to the infantry but have increased logistics and, recovery and repair capacity.

The light cavalry is equipped with a Light 4 x 4 vehicle mounting an NLOS-guided weapon and/or 40mm GMG. The exact type is less important than:

- Crew of three
- Drive 800km
- <4 tonnes GVW
- Able to fit inside a 20foot ISO
- Mount the relevant communications, sensor and weapon pack.

The light cavalry requirement is for a unit that can move at least 800km across and is comfortable operating in arid terrain with little infrastructure, which would logistically constrain traditional armoured units. This means a minimum of 72 hours without resupply. The entire unit should be configured to deploy by air within wide-body airliners or military transports, so all chassis should essentially conform to the specifications of the fighting vehicle. This is not technically demanding and is within mature technology.

The Low-Level Air Defence Group would be equipped mainly with MANPADS, such as RBS-70 of Starstreak LMM, with the possible addition of a 30mm cannon.

It should be noted that during some of the simulation and DSTL licenced wargaming to support this work, an experimental vehicle mounted an M230LF 30mm cannon combined with a Javelin missile in a remote weapon station. This is essentially the same weapon used on the AH-64 Apache. The simulation showed the system to be highly effective even though the High Explosive Dual purpose 30mm x113 ammunition modelled only perforated 25mm of RHA compared to the nearly double possible with 40mm. The 30mm cannons that usually equip NATO and Western armoured vehicles are 30mm x 173, which use APDS ammunition capable of perforating at least 100mm of RHA.

However, this is a substantially heavier, more expensive and complex weapon.

From a UK perspective, the Light Cavalry would mean the capability to drive from southern England to Warsaw in just over 48 hours, equipped with 40 ATGW posts plus at least six rounds per post. This represents a very viable anti-tank force. There is a strong evidential basis for this assertion in the 2002 RAND Study, "Lighting Over Water." Based on simulation experiments, this study concluded that a rapidly deployable US light force could effectively counter an enemy armoured formation by employing NLOS or fibre optically guided weapons, which are now commonplace and would equip a light cavalry unit. From an Israeli perspective it would be a screening force able to deploy deep into desert areas and be sustained using support helicopters for re-supply. As of 2023 both the UK and Israel have units broadly like what is described, and indeed, the UK units are called Light Cavalry Regiments. The Israelis have several specialist sub-units mounted in light vehicles.

Armour and enclosure are desirable, and the French Arquus VBL Mk3 would demonstrate this. However, with every level of protection comes an increase in gross vehicle weight, thus commensurate increases in cost and force structure sustainment.

With increased weight comes an increase in signature. Thus, the Thales Hawkei might be deemed a sound choice, but that comes at 7-10 tonnes and $1.3 million. The Light Cavalry must be a true alternative to the heavy. Ultimately, this is not about levels of protection but degrees of performance for a constrained set of resources.

The heavy cavalry is completely different, as in being equipped with tracked armoured vehicles armed with 120mm or 105mm main guns. The CV-90/120 would be such a capability. We previously discussed this as MBT-L in Chapter 2. The air defence variant would mount a 30 or 35-mm equipped air defence turret with missiles. The overall vehicle criteria would be:

- Crew of three
- Drive 800km
- <35 tonnes GVW
- Protection level >STANAG 4560 Level 6
- Active protection system (ideally capable against loitering munitions)
- Mount the relevant communications, sensor and weapon pack.

The 800km requirement would mean that all vehicles would need rubber tracks. It must be conceded that it is extremely unlikely that an entire unit of this type could complete an 800km unit move in 12 hours, but the capability of a single vehicle to demonstrate this has beneficial consequences in proving sustained manoeuvre over long distances.

These vehicles would represent the single greatest equipment investment in terms of AFVs that the Monash-type force could make. In the eye of the beholder, they would look like tanks, but any idea of employing them as tanks would need to be strongly resisted. To emphasise this, there is no requirement for the heavy cavalry to have any minefield-breaching capability or earth-moving equipment.

The heavy cavalry should possess sensors and communications far more capable at the vehicle level than the light cavalry. Their armour should primarily concern artillery fragmentation with active protection systems dealing with guided weapons, RPGs and loitering munitions.

The defining quality of this vehicle choice or type is its manoeuvre capacity in that, being tracked, it will cross ground which wheeled vehicles cannot.

Conclusion

We started this chapter by exploring the traditional ideal of what cavalry reconnaissance might or should be but faced with the constrained model and the need to make clear doctrinal choices forcing a distinction. The suggestion was that the Cavalry were not distinct because they performed reconnaissance but because they operated in open terrain and were optimised to do so. Cavalry operates primarily as a mounted force, limiting their effectiveness to the ground, allowing them to exploit their mobility and firepower. However, the key insight is the need to destroy and defeat the enemy reconnaissance and threaten or offend any enemy they may encounter. Heavy cavalry, therefore, have greater offensive potential than light cavalry.

Endnotes

1 I can stand by this claim merely because I have been on first-name terms with at least five commanding officers from various UK MBT units and about ten sub-unit commanders.
2 Naylor, Sean, *Not a Good Day to Die*, Berkley Books 2005.

7

FIRES

Artillery has been central to modern warfare since the late 19th century but has not occurred in isolation from air surveillance, improved cartography, and even naval gun technology. Thus, artillery is a vast and important subject.

It is important to recognise that artillery and its associated technologies, termed "Fires", are the critical enabling factor on which almost all else relies. This chapter will split fires into two distinct parts.

The first part of the following section will focus on 155mm guns and the close support role.

We should avoid a debate about mass versus precision at the conceptual level. As we saw with tracks and wheels, there is not a fruitful discussion coming down hard on one side or the other. Much of the problem with that debate is the lack of open-source data or data points everyone can agree on, but sensible insights can be gained.

The second part is deep fires, which, for this work, are the equipment, organisation, and training that allow the Division to degrade, disrupt, and attrit the enemy before the close battle, be that offensive or defensive. Understanding this as a separate and distinct action within the Divisional framework is critical. To assume that all the separated parts of the surveillance, intelligence and target finding can be grouped to be centrally controlled and then allocated on an as-needed basis attracts unwanted complexity and friction. The pop concept of any sensor, any shooter, is fundamentally flawed. Much of the effort here is directly associated with gaining the freedom of air action discussed in Chapter 3.

Close Support

How many 155mm shells must you land in one hectare in one minute to gain the required effect against the target?

Conventional artillery doctrine seeks three distinct effects: suppression, neutralisation, and destruction. Suppression and destruction in this context have distinct meanings, unconnected to the same words used to describe attacking air defence systems.

The important point about all such distinctions is that they are effects generated against human beings regarding killing or wounding. Killing or wounding one man might suppress all the others in the vicinity.

Suppression is the weight of fire that prevents any activity on the position when rounds are landing. Neutralisation is a heavier weight of fire that produces a shock and casualties to the degree that no effective action is taken for some minutes after the firing.

However, as with suppression, the effect is temporary. Destruction is the weight of fire that produces so many casualties that all forms of resistance are impossible. Based on that, it is obvious that the weight of fire needed to destroy four trucks parked in the open would be very different from that needed to kill or wound soldiers in trenches with 1.2m of overhead cover in the same location or four main battle tanks. The problem is that none of the specified effects are absolute, and in war, there is no way to ascertain if those effects have been achieved until your forces arrive at the position. The fire used with the intent of suppression may well cause some destruction. Fire intended to destroy might only have been accurate enough to produce some suppression and neutralisation.

Thus, the effectiveness of 155mm in this discussion is about the number of rounds landed within an area over time.

The CEP for a US M549A1 shell fired at 25km is 146m based on category one target location.[1] One hectare equates to 100m x 100m. The CEP is the circular error of probability is the area in which 50 percent of the shells fall, thus it seems likely that the total effects footprint of M459A1 at 25km is close to 300m or approximately 7 hectares. The British Royal Artillery's weight of fire tables are based on a 200m x 200m area of four hectares.[2] Thus, not every hectare will receive the same number of shells because artillery rounds from a single gun are dispersed in an ellipse because most of the dispersion is range, not windage. Dispersion increases with range, so a 146m CEP at 25km is notionally >200m at 30km.[3]

The one-minute time factor is a function of how quickly guns must be displaced from their firing position, given an enemy capable of conducting competent counter-battery fire. That is the exam question which drives all else. It is fair to insist on a qualification regarding the required effect and the nature of the target. There are also the added shell and fuze-type dimensions, plus how the trajectory affects the splinter footprint. No one can claim that artillery is simple.

Do you need to displace after one minute? Digitised fire control systems mean that guns can now disperse into single- or two-gun detachments, greatly reducing the threat from counter-battery fire. Guns only need to displace >500m in any direction to mitigate the effects of gun rounds being fired on their detected location. Assuming detection by counter-battery radar, it takes time to relay firing data to guns, the guns to lay, and the shell's time of flight, which at 20km will be over 1 minute, it seems reasonable to assume guns need to move within 3 mins of firing. For this discussion, the best practice is firing 12 rounds and displacing more than 500m.[4] This means that displacement should be achieved within 3 minutes as a baseline training standard from when the first round is fired.

The critical observation here is that the weights of fire need to be validated by instrumented field trials against real target arrays, such as a dug-in Russian-type BMP platoon. This platoon has a laydown of 400m frontage and 300m depth, or 12 hectares.

Artillery will need to suppress or neutralise the front edge of that position, where most of the direct-fire weapons will be concentrated. That means suppressing 400m of frontage. The maximum number of rounds a unit of 18 guns can fire in two minutes is 216, so given the parameters discussed so far and having one artillery unit support one sub-unit attack against a platoon, any required effect must be achieved within 216 rounds.

Concentrating fires at range so that all the rounds fired land within the target area is the assured way to reduce the expenditure of rounds, thus the times the guns are exposed, and increase the effect of the target. Two general approaches currently exist.

The first is precision-guided artillery rounds such as Excalibur. Excalibur costs $100,000 per round compared to $5,500-6,000 for a complete unguided round. Given Category One target data, shells can destroy individual positions so that the BMP platoon can be destroyed by as little as 3-5 rounds. Thus, the cost is $3-500,000, which equates to 90-166 unguided shells, depending on the exact cost.

The second is precision guidance kits (PGK), which are GPS and INS fuses which screw into the nose of any NATO standard 155mm. The same system is available for 120mm mortars. PGKs reduce the shell's dispersion to 50m at any range. Given the lethal radius of 80m, it might be expected the two PGK rounds will generate a significant effect against any target within 50m of the target datum. PGKs are reported to cost about $15-20,000, so the total cost of a 155mm PGK shell is about $26,000 or a ¼-⅙ of Excalibur. Conversely, that equates to 4-7 unguided shells.

"Vulcano" is another precision round. It is a 127mm projectile fired from a 155mm sabot and can travel up to 70km from an L/52 calibre gun. Excalibur's range from an L/52 calibre gun is 50km. 127mm round lacks the terminal effect of 155mm.

The real-world problem is that artillery often needs to target an area, such as a forest or a reverse slope. Precision guidance is highly effective when precise target data is available from UAS or ground observers but not so useful when seeking to affect an enemy for which you only have an approximate location. Precision is a tool, only appropriate to some conditions, so it does not remove the need for massed area fires. One added benefit of precision is the ability to use 155mm in close support with a reduced risk to troops close to the target, which has substantial and obvious benefits. Gun batteries could well consider organising around having one gun dedicated to Excalibur employment, leaving the other 5 or 7 able to employ PGK as and when required.

The last 10-20 years have also seen 155mm ammunition types proliferate. Beyond smoke and flares, the two notable types are base ejecting sensor fused munitions such as BONUS and SMART and pre-fragmented HE payloads known as "Super HE." Sensor-fused munitions are designed to target armoured vehicles, including self-propelled guns.

The intention is that one round eject two sensor-fused munitions, which then search a circular footprint of about 200m for a target, which, when detected, initiates an explosively formed penetrator to strike the target. The effects could be substantial when combined with PGK against armour moving on a known route. The limitation of such rounds is their cost, which is reportedly close to, if not equivalent to, Excalibur, meaning only a very small number will be available within a battery.

Super HE, which is currently only in service with Israel, is a shell whose base ejects a cylinder of machined fragments with an explosive core. A drogue stabilises the cylinder vertically to detonate at an optimum height above the ground, producing a perfectly circular fragmentation pattern instead of a normal 155mm butterfly footprint. Current industry estimates state that it is 40 percent more effective than conventional rounds.

Mention should also be made of so-called cluster munitions, such as rounds that base eject bomblets, anti-armour bomblets, or various types, most of which have both a shaped charge and fragmentation body. Current national treaty limitations, such as the Convention on Cluster Munitions, will decide the consideration of these types. Still, notably, China, Russia, Iran and North Korea deem them effective, as does Israel, Ukraine, the US, Estonia and Latvia. The debate is not about military effectiveness but rather the percentage of unexploded munitions, which present a risk to civilians post-conflict as well as their troops arriving on or passing through the objective. A M864 155mm round ejects 72 submunitions, meaning that 100 shells would dispense 7,200 submunitions, meaning that a 1 percent dud rate equates to 72 rounds. A 0.2 percent dud rate still equates to 14 submunitions. Cluster munitions are substantially more expensive than high-explosive rounds. The issue of when and if to use cluster munitions must be considered against the same balance of investment choices that have driven the rest of this discussion and that their use would be as a percentage of the conventional rounds used in a fire mission.

The issue is what design of 155mm howitzer, be that towed, tracked, self-propelled, or truck-mounted, exactly conforms to cost, weight, and complexity. Complexity strongly leans against reliability, so while fully automated burst fire 155mm is a reality, there are attendant costs of ownership and maintenance burdens to consider. Gun systems incapable of manual reversion in loading and firing are a significant liability unless the number of guns is such that the time between failures mitigates that risk.

With some exceptions, most artillery converges on L/52 calibre NATO JMBOU (joint ballistic memorandum of understanding) compliant guns. However, the focus of the discussion on guns is misplaced when an L/52mm barrel is essentially agnostic of the shell placed in the chamber and the effort taken to get it there. In poetic terms, "the gun doesn't care."

While it is technically possible to operate a 155mm self-propelled self-loading gun with just two crew members, the required administrative, maintenance, and redundancy debt inherent to field artillery organisation means there is no actual cost savings in manpower. Most aspects of crewing or supporting guns are extremely physically demanding and require high concentration levels.

Every gun system currently in existence demands that the shell be manhandled from the pallet to the platform or resupply vehicle at some point, and this needs to be done quickly and repeatedly. The gun crews all demand very high standards of strength and physical fitness. A hard 12 hours of operation might see a gun crew complete 40 fire missions, consuming more than 400 rounds. That effort can only be sustained before barrels with a notional life of anywhere between 7-2,000 effective full charge rounds must be inspected to ascertain their condition or swapped out completely.

The most important realisation is that whatever gun system is selected, the cost of stockpiled ammunition far outstrips the cost of the gun. At a minimum, an Army needs to hold 30 days' stock of 350 rounds per day per gun. 10,500 rounds per gun at a possible cost of $6,000 per round is $63 million per gun. Even at $3,000 per round, the figure is still $31.5 million.

Artillery logistics is critical enough to be a specialisation separated from normal combat service support functions, requiring dedicated trucks, crews, and handling equipment. Four hundred rounds per day per gun is not beyond reason. In 1988, the British Army's Staff Officer Handbook stated that 155mm expenditure would be 360 rounds per gun per day (RGD) for M-109 and 480 for FH-70 towed guns[5]. The daily expenditure rate is axiomatically the rate at which your logistic system needs to support and the ammunition hold you need to lift or have immediately to hand. This means a unit of 18 guns would have 38-42 x 18-tonne trucks of unit mobile stocks.

There is thus a strong argument for bringing artillery logistics within the artillery arm as a specialisation. The "why now" answer is that gun range and dispersion require far more granular control of logistic support. Ground dumping shells into caches or hides from which guns come to "feed" and resupply should not be discounted given adequate security. What logistics cannot do is concentrate to the degree they become a detectable target. Logistics is the Achilles heel of artillery as a critical function requiring skilled application.

Also lacking in appreciation is that the same communications architecture that ensures you can use UAS is indivisible from that used to control, manoeuvre, and sustain guns and vice versa. The communications fit in the battery command post and on the gun platforms, and the attendant digital fire control applications may be more important than the type of 155mm gun they are fitted to.

The Monash Close Support Unit

The Monash Close Support unit had 18 x 155mm guns. This maximum can be sustained within the 100-vehicle platform limit, allowing each gun to be supported by two EPLS (Enhanced Palletised Loading System) ammunition resupply trucks holding 170 rounds. Including the rounds carried on the gun, more than 360 rounds should be immediately available.

For simplicity, cost, and reliability, the guns should use a wheeled chassis and be manually breech-fed. This will demand a crew of at least five, so battery manning will be at least 30 men on the guns alone. Given that each gun is supported by two trucks, each with a three-man cab, there is potentially a "ghost crew" of another four men to account for the arduous nature of working on the guns. Guns should be L/52 NATO MOBU (memorandum of ballistic understanding) compliant. One gun per battery would be designated as the "Precision Gun, " meaning it will hold the relevant small number of precision-guided rounds needed to address targets suited to their employment.

The battery of the sub-unit support detachment is roughly the same as the infantry, as do the medical and repair sections, albeit slightly altered to suit the equipment and organisation. For example, given a truck-based gun platform, barrels should be able to be replaced without the gun leaving the battery.

Notably, the Close Support Unit has no organic sensors but is optimised to communicate with the Artillery Forward Observers found in all Infantry and Cavalry Sub-unit HQs.

Deep Fires

Deep fires have existed since WW1, and the Royal Artillery placed considerable effort into creating this capability via observation balloons and railway guns capable of reaching ranges of 19 to 30km.

Deep fires are materially and resource-intensive, requiring careful allocation within an economy of force framework. Almost anything targeted within the relative area of operation can be replaced or has some form of redundancy, including enemy commanders and staff. It is comparatively simple to assemble and service a high-value target list. It is less simple to realise the tangible benefits of doing so unless it creates the possibility of gaining some further advantage that allows you either to defeat the enemy's offensive action or conduct your offensive action. This is not to say attrition does not have material benefits, but even then, striking high-value targets needs to be predicated on the aim that doing so will allow the detection of other high-value targets. As discussed in Chapter Three, the entire purpose of the deep battle is to render the enemy less well-prepared for close and decisive operations. It is not a geographic concept despite range being a key component. Thus, the conduct of the deep battle must balance the resources needed for adding friction and fear to the enemy's everyday operation and the conduct of activity directly related to decisive ground manoeuvre and/or massed fires.

All of this must be balanced against the need to protect the conduct of the deep battle against enemy interference. The additional burden is the requirement for 24-hour conduct, meaning staff, sensors, and effectors must be manned by 2-3 shifts to ensure the capacity for continuous engagement.

Suppression and Destruction of Enemy Air Defence and C2 (SEAD/DEAD)

In practical terms, the division of labour between SEAD/DEAD and deep fires should not exist. They are the same. They should merely be distinctions in targeting sequence, but their value cannot be overstated. Given the previously mentioned requirement to create freedom of air and electromagnetic spectrum action, this is probably one of the most important combat support tasks emerging or extant on the modern battlefield. The idea that the destruction or suppression of enemy air defence is an air force-only task is not something history supports, at least as concerns Israel and its procurement of MGM-52 Lance ballistic missile with cluster munition warhead, its previous use of ground-launched anti-radiation missiles and then its development of the Harpy and Harop loitering munitions in the 1980s and 1990s was expressly for killing enemy air defence assets. This is an area based on considerable user experience.

Land force DEAD/SEAD is range-limited compared to air force ambitions and requirements, but the mission it needs to fulfil is also distinct from that which the air forces normally aim to achieve. Land force DEAD/SEAD is designed to create freedom of action for those air assets, manned or unmanned, that support the reconnaissance and fire missions needed to attack the enemy before their involvement in the close battle. This means that the Land force DEAD/SEAD needs to be planned, controlled and coordinated at least at the division and maybe corps level. It should mainly target the enemy air defence systems within their Division or Corps battlespaces. This means the potential requirement to strike targets 1-200km forward of the forward line of own troops and thus possessing or accessing the ISTAR assets that can detect and geolocate remitters at that range. This will be a battlespace-defined mission requiring extensive electronic and electro-optical reconnaissance, some of which may require joint-service or allied cooperation. The mission can be achieved in its simplest form by loitering munitions autonomously attacking detected emitters based on a pre-programmed threat library.

This is a mature technology but, in isolation, will only target active radars within the search area. Suppose the enemy air defence system relies on targeting data supplied by radars outside the target area or uses optical engagement. In that case, the loitering munitions are unlikely to attack. This means some activity must force the enemy to unmask, use radar, or engage the loitering munitions or UAS. In its ideal form, the DEAD/SEAD mission means targeting and killing the widest range of radars and emitters that time, resources, and technology will allow.

While autonomous loitering munitions are self-contained sensors and effectors, other systems require active control links. This is critical to resources because competence in this area has far wider advantages. Success should produce a combined and synergistic effect of degrading the enemy C3I and air defence to the degree that creates freedom of action, which can be exploited in terms of fire and manoeuvre to detect and attack further targets in line with the concept of operation described in chapter three. It is important to realise that the aim of the wider activity is not just targeting the enemy systems themselves but using the system signature to identify opportunities to target enemy commanders and staff, air defence personnel and UAS operators. Killing radar operators is substantially more consequential than just rendering the radar inoperative. Warfare is a human activity, so breaking human will is paramount.

The ability to detect and target high-value targets at a distance is a critical advantage which needs substantial investment.

Unmanned Air Systems

UAS has been a significant part of all Western armies' inventory for over 40 years. In the Vietnam War, the US flew jet-powered long-range UAS extensively. The more common piston engine, the small UAS common today, was active in the 1982 Lebanon War and has been in most conflicts since. Despite media claims to the contrary, little new or novel has emerged in recent years. The use of airborne camera feed to detect targets and control fires, which has existed since the 1980s, is still the primary model for employment. Gradual evolution has produced a reduced operator, support and sustainment burden required to employ such systems. In some parts, this is a direct function of the reduced cost of miniaturised sensors and control feeds. This has commensurately lowered the barrier for entry to UAS operations, meaning that capable systems are within reach of less developed armies and non-state actors. Despite all this, UAS of every sort is still prone to the collective misunderstanding that prevents their effective employment. UAS are just a means of positioning a sensor. 80 percent of the cost of most small to medium military standard UAS will be its sensor and communications link, not its engine and airframe.

Thus, discussing a UAS requirement should start with the sensor performance required to achieve the mission or task required. The most important function of the sensor is the ability to detect and identify targets at a range commensurate with providing target location data. That can be as simple as the operator using a map, and a 250-350g payload can now generate Category 3 target data based entirely on slant range GPS calculations. The same payload is thermal imaging capable and can detect walking men at a 1,000m slant range. Military standard quad-copter type UAS using encrypted SDR communications links can operate at 10km for 70 minutes or more up to a pressure altitude of 12,000ft. This UAS can be carried in a backpack and operated by 2-3 men.

Smaller versions can be operated by one man but are limited to 5km and about half the endurance. The obvious conclusion for some is to demand that every platoon, sub-unit or specialist detachment has a UAS system; however, 5km, let alone 10, far outstrips the area of responsibility that may ever be allotted to a platoon or sub-unit. The capacity to look over a wall or beyond the next wood or tree line might seem immensely useful.

Still, the primary purpose of UAS is to detect targets for indirect fire systems and, payload-dependent, gather electronic intelligence. The UAS should also provide a node on the communications network. In terms of infantry, this would best place small UAS with mortar and NLOS-capable in anti-tank platoons or detachments. Still, even small UAS come with some support debt concerning dedicated operators, equipment support, battery charging and potential span of command issues.

Elevating the same logic to the formation level, the UAS requirement would be to find targets for and control the fire of 155mm howitzers and rocket launch systems. Given the current potential of 155mm to fire out 70km, using specialised ammunition natures, this requires payloads capable of generating category 1-3 target grids and even laser designating. This means a 9-12kg payload capacity, which axiomatically means a fixed-wing airframe of around 50-55kg able to operate out to 150km for 24 hours. 9-12kg payloads can detect NATO standard targets at 15km. Current technology allows 50-55kg or heavier airframes to have vertical take-off and landing capability. Such a UAS requires no more than a 4x4 vehicle and two crew. That said, 24-hour operation probably demands a minimum of 4-6 crew. The same types of airframes often possess the capacity to carry electronic intelligence (ELINT) payloads and even small synthetic aperture radar. The technological capacity of these systems far outstrips most land force requirements. Much can be simplified if the focus is directed towards finding targets. Giving one airframe multiple sensors can lead to a division of attention as the best time and place for an ELINT payload to detect targets may not be the same as that of an electro-optical payload. Suppose we assume that as regards equipment, organisation and training, UAS detachments can be 1-2 vehicles and 3-6 men allocated to artillery sub-units or batteries. In that case, UAS detachments can be both simple and effective. The same logic would apply to electronic warfare units. In 2024, most military-grade UAS of the type discussed cannot be disabled by GPS denial. The danger or vulnerability of a UAS control link is identical to that of any SDR IP-capable radio, and unless the control link is hi-jacked, correctly equipped UAS will revert to a pre-programmed inertial recovery sequence or automatically select and log onto an alternate communications means. UAS can be launched electronically silent and proceed to a search area, where they will only initiate communications based on their sensors' image recognition library. This is a mature technology and in current service.

SAM systems and anti-aircraft artillery are the overarching threat to the formation and divisional UAS. Finding and killing such systems and their operators is thus the primary enabler to gaining freedom of action for all UAS.

Tethered UAS, systems connected by a physical link to a vehicle or ground station, need to be mentioned. The cable normally carries the downlink. Fibre optics give these systems considerable potential. Completely electronically silent, they are immune to EW but are limited by high winds more than free flight systems. However, the potential of positioning a 3kg sensor at 100m altitude should be considered.

Large Air Force UAS, as in systems requiring runways operated from fixed facilities, can carry electro-optical sensors capable of detecting and identifying military vehicles and air defence systems 70km slant ranges and generally fall into the High Altitude and Long Endurance (HALE) description of UAS so will initially be tasked to missions supporting Air Force SEAD/DEAD and counter-air. They fall outside the area discussed in this work.

Loitering Munitions

First entering service in the early 1990s, loitering munitions or "Kamikaze" or "suicide" drones are distinct from normal UAS in several ways. Firstly, they carry a warhead. Secondly, they are optimised to dive and strike a target. Thirdly, they are designed for hunting for a target via human direction or automated means rather than strike a pre-programmed target. It would qualify as a cruise missile if designed to strike a pre-programmed target. Until recently, loitering munitions were not intended to be recovered if not able to find something a target. However, the cost-benefit of that issue has led to at least one variant being equipped with parachute recovery.

The Lancet-3 Russian loitering munition has reportedly gained many kills against Ukrainian armour, artillery, and air defence, but, notably, most Lancets are engaging targets already detected by UAS. The countermeasure that has a disproportionate effect is to be undetectable from the air and has proved effective since 1915, or even before.

Loitering munitions appear conceptually simple to operate for those without practical experience in their employment. The actual issue of their employment, not usually apparent in most discussions, is their integration into the overall fire plan intended to enable or defeat a scheme of manoeuvre. This means loitering munitions must be controlled by the same level of command that controls artillery units tasked with deep and counter-battery fires.

The capability itself is undeniably simple. Once launched, the loitering munition flies to the pre-planned search area to find the target, usually via an electro-optical sensor. Once detected, it closes with and dives onto the target. Depending on the size and type of warhead carried, it can destroy most targets, from a main battle tank to an air defence radar system. It can also target specific rooms or building areas, but the warheads are generally not optimised for that role.

The last decade has seen the development of specialised anti-personnel loitering munitions designed to hunt targets inside buildings, but this is generally a special forces-type capability, so it falls outside this discussion. As with UAS, loitering munitions cover a vast size, weight, and capability spectrum. This spans munitions that can be launched and controlled by one man. With a weight of 4kg, they can fly for 40 minutes out to a range of about 15km with a 500g warhead. This then scales up with commensurate increases in range and payload. Table 7.1 is a notional scale of loitering munitions performance assembled from various sources.

Item	Weight	Warhead	Range	Endurance
LM1	4kg	500g	15km	40 mins
LM2	9kg	1.5kg	40km	45 mins
LM3	14.5kg	4.5kg	60km	60 mins
LM4	50kg	8kg	150km	120 mins
Zala Lancet-3	12kg	3kg	40km	40 mins

Table 7.1 Notional Scale of Loitering Munitions

More capable systems exist, but in their simplest form, loitering munitions depend on a control link and, thus, an emitter. As with UAS, the security of that link will be critical to the system's employment. The focus of this debate is the use of loitering munitions and for what end. In the context of this work, the discussion is how to better prepare armies or land forces for warfare. If the enemy cannot contest the air and EM spectrum, loitering munitions have the potential to be decisive in terms of battles and engagements. Even if the enemy possesses low-level air defence in the shape of HMGs and MANPADs, an inability to detect and degrade the control link will offer them little defence. A well-equipped enemy with radar and electrooptically controlled missiles of gun air defence will be able to shoot down loitering munitions, except, as described earlier, the origin of the loitering munitions in the shape of the Israeli Harpy was an anti-radar seeking autonomous loitering munition, several which could be launched into an area to target radars and emitters. The arguments for or against Loitering munitions revert to a cost, weight and complexity matrix regarding a balance of investment. Cost is an issue. As of 2023 a loitering munition capable of killing an MBT (LM2) is more expensive than the comparable anti-tank system. Based on the value of sales declarations made by various governments as of 2024, an LM2-type munition may cost $50,000-100,000 or even more.[6] Given this cost, it makes sense to specify that loitering munitions should be recoverable in case of not detecting a target, but this may be impractical for autonomous loitering munitions operating in the anti-radar role. There is also the factor that in a general war setting, it is unlikely that any loitering munition will be unable to find a target of some sort if the desired value of the target has not been acquired. Recovery is ideal but maybe not the exclusion of more useful types.

The loitering munitions advantage is that detected unit or sub-unit level armoured manoeuvre can be defeated at range, assuming that the

MBTs do not possess some active protection system capable of defeating the incoming munition. What applies to MBTs also applies to IFVs, APCs and artillery. NLOS Anti-tank-guided weapons cannot perform a counter-battery task at any useful distance. Loitering munitions can, but that may not be their best form of employment. There is a danger in overstating loitering munition's potential, especially given they have existed in varying forms for over 30 years. Perceived low cost, weight, and ease of employment seem to drive wider acceptance. Loitering munitions cost substantially more than artillery shells, mortar bombs and most common rocket types, and based on current data, they are not substantially cheaper than NLOS-capable guided weapons. Loitering munitions are still relatively less effective in forests and urban environments.

Suppose the electronic freedom of action exists to operate loitering munitions. In that case, the same is true for UAS with better target detection potential, far higher-resolution sensors, and more robust command links. At the application level, there would seem to be a danger in creating what simplistic concepts call a sensor-to-shooter loop; with the loitering munition, the sensor is killed while killing the target, thus no longer of use.

By far, the greatest potential possessed by loitering munitions is related to their origin, where they can autonomously attack and kill an air defence and electronic warfare networks using anti-radiation and signature recognition seekers, thus creating freedom of action of UAS. Thus, the balance of investment for "suicide drones" is not to target AFVs or troops but to create the conditions for which UAS employment becomes viable.

Electronic Warfare and Cyber

As should now be obvious, electronic warfare or CEMA (cyber and electromagnetic activities) is critical to modern combat support functions, even as far back as WW1 and WW2. These functions of both EW and cyber are essentially identical in that;

1. You can intercept and interpret electronic signals or communications.
2. You can jam, disrupt, and degrade electronic signals or communications.
3. You can take control of part or all of the enemy communications network disrupting, degrading and denying effective command and control while the enemy is unaware this is ongoing.

The above description removes a lot of the mystique from CEMA, and the potential of cyber has been vastly overstated compared to its proven effects. The point that seems routinely ignored is the human dimension critical to CEMA. To make a relevant digression based on personal experience, in 2016 I had dinner with an Israeli friend who had just left the Army as a senior officer. He had an extensive human intelligence background, and I asked him what he was currently up to. He replied, "Cyber", which made me ask why when he had no background in the area, what could he offer. His simple answer was that every keyboard demands human fingers, and every algorithm starts with a human author. His interest in military EW and cyber was finding and killing the operators, not contesting the spectrum. To his mind, electronic warfare was not a spectrum-based competition but a physically lethal one where you sought out the humans in the loop, making the loop irrelevant. Any EW asset on the battlefield should be located and killed, and this is coherent with the need to establish air and EMS freedom of action.

Regarding the concept advanced in Chapter 3, EW/CEMA should focus on finding targets or gathering intelligence to support strikes commensurate with the relevant scheme of manoeuvre. That is not to suggest that degrading or denying enemy communications will not be important, but where possible, hitting a transmitter with a warhead is more effective than jamming the signal from it. Logically, this all reverts to a simple mission statement that the primary role of EW/CEMA is to create the freedom of action for the air and electromagnetic spectrum. The number of platforms, troops, means, and required methods mostly fall into the classified domain, and there is no such thing as low-cost EW/CEMA. Still, resources can be better understood if directed towards one coherent mission set. The cost and complexity are reduced substantially if the mission set is confined to locating and killing emitters.

Rockets

All we have discussed with guns can mostly be applied to rockets, but rockets should not be seen as an alternative to guns in the same way violins should not be considered alternatives to trumpets. However, their effect on targets should be understood as the same thing as a salvo of 122mm rockets can generate effects identical to those of 155mm as regards suppression, neutralisation, and destruction. Basic artillery rockets can deliver a very heavy weight of fire in a very short time. Far more than guns are capable of from a lesser number of platforms. Modern guided rocket systems can generally engage at longer ranges more accurately, with greater effect and from lighter platforms than gun systems. The problem with rockets is volume, weight and rate of fire, but the reality and the potential of MLRS systems need to be understood as part of Divisional or Corps-level combat support.

Rockets generally require only 4-18 tonne truck launch platforms, which are cheap and easy to maintain. Depending on the requirement, considerably smaller platforms can be adapted to launch rockets. An HMMWV-type vehicle can carry and launch 8 x 122mm rockets. A Land Rover or similar vehicle can carry 4 x laser or GPS-guided 170mm rockets with a range of 36km and a 14kg warhead. Rocket platforms can also be displaced after firing far quicker than most gun systems.

Because most modular rocket systems use pods of rockets, reloading time, depending on the specifics of the platform, is usually less than 10 mins. 155mm gun that might need 36 rounds of manual restocking on the platform and has a crew of 5 would require each crew member to lift and carry 7-8 45kg rounds each, so not a task likely to be accomplished in more than 5 to 8 minutes. Crewing a rocket platform is nowhere near as physically demanding as crewing a gun, and rocket platforms can reliably sustain themselves with 2-3 crew. Given modern digital fire control systems, rocket targeting and launching are simpler than guns.

The real potential of rockets is "massed precision". One vehicle can launch 36 x 122mm INS/GPS rockets to a range of 40km with a warhead of 20kg. Each rocket can hit within 20m (10m CEP) of 36 separate aim points, all impacting within 1 minute of each other. This is based on GPS or INS trajectory correction. What i's possible for a 36 x 122mm is the same for a 12 x 227mm rocket at 70km with 90kg warheads. Theoretically, it is possible to neutralise or destroy one enemy platoon for every launch vehicle available to support the action.

Rockets have additional advantages inherent to their design. For example, while a 155mm shell can carry two sensor-fused munitions, a 227mm rocket can carry four. This means one rocket platform can launch 48 SFMs instead of eight guns firing six rounds in one minute to achieve the same effect.

The other perhaps more obvious issue for rockets is long-range precision targeting or high-value assets; thus, rockets are the primary means by which the Division or Corps can kill enemy air defence and C3I. A 370mm rocket system with a CEP of 10m and warhead of 140kg can strike within 300km, and it is a decade-old or more proven technology. One launcher can fire four such rockets.

Rockets and launch systems are light engineering compared to 155m shells and guns, which require boring barrels, forging, casting and handling molten metals. Rockets can be made in any warehouse, assuming access to the correct tools and jigs. The Hamas terrorist organisation manufactured many thousands of rockets with no heavy engineering tooling at all. The level of infrastructure manufacturing for rockets is considerably less than for conventional artillery and shells. However, manufacturing guided rockets will generally have a higher unit cost than shells. The cost comparison becomes less acute when compared to guided 155mm.

Perhaps the major conceptual problem with rockets is the purgatory of choice. There are very few indirect fire missions that rockets cannot address in some way, but to address each well requires a different rocket system, thus holding several different rockets ready to fire or stored. Rockets can be employed in close support tasks, but the best use of weight and volume within a constrained logistic system means rockets are best employed in support of the deep battle rather than the close.

The Deep Fires Unit

The Deep Fires unit is nearly identical to the Close Support unit but with the following differences.

Three batteries comprise two of five multi-barrel rocket launcher (MBRL) platforms and one of eight loitering munition launch platforms. The MBRL batteries each have 10 EPLS 8x8 vehicles with trailers, each carrying 4 rocket pods. The loitering munition Batteries have ten resupply vehicles carrying 30-40 rounds each.

The MBRLs are truck-based and carry two rocket pods. Each battery employs 306mm rockets with 120kg warheads, each with a 30-150km range and four rockets per pod. The time of flight to the maximum range is less than five minutes.

The loitering munitions carry a 30kg warhead and can fly out to 150km or more with a 6-hour endurance. They can be equipped with either an electro-optical seeker or an anti-radiation seeker. The level of autonomy in each case will be ROE-dependent, but the battery has eight control stations for human-in-the-loop targeting. All munitions are rail-launched rather than canister-launched to save weight, cost and complexity.

In all other respects, the organisation and equipment parallel the Close Support Unit.

The Deep Targets Unit

The Deep Targets unit is also organised broadly similar to the previous two units, with three batteries and their supporting detachments.

There is a UAS Battery, a Multi-mode Radar Battery and an Electronic Warfare battery. All of these report to a Targets Group housed in 4 vehicles that work to create targeting information from the various sensor inputs and then direct and control the fires of the Deep Fires unit. The Targets Group within the Deep Targets Unit will run the deep battle under the direction of the General Officer Commanding.

The UAS Battery comprises 30 UAS detachments, each employing one vehicle and three crew members, with three airframes. The UAS is a fixed-wing airframe weighing around 50-55kg that can operate out to 150km for 24 hours. 9-12kg payloads can detect NATO standard targets at 15km. It takes off and lands vertically. It can operate in a GPS-denied environment. Assuming 2 Brigades in the Division, each brigade is allocated ten UAS detachments with electro-optical payloads. Ten are retained by the unit with ELINT payloads to work with the EW battery. The EW battery focuses on locating emitters for targeting rather than jamming or interception. The Ten vehicles equate to five detachments.

The Multi-Mode Radar Battery is 10 radar-equipped vehicles, and the equipment capability for those radars will be specified in the air defence chapter.

Conclusion

Fires win battles and engagements as the most decisive units available within the division or corps. Infantry and fires are the best choices if you can only fund two unit types. There are little fires that cannot strike or destroy, so funding should sustain that intent along with the necessary means to locate targets and direct fires.[7]

Endnotes

1   Lt Col Mike Milner Combat ammunition project officer. Excalibur capability brief 2012.

2 In the British Army as of 2023, Weight of Fires tables are classified. The published data in Staff Tables is for exercise purposes and logistics planning.

3 Lt Col Mike Milner Combat ammunition project officer. Excalibur capability brief 2012.

4 Amongst Royal Artillery Officers. Opinions varied widely as to the required distance, also noting that displacement offered no advantage if being tracked by UAS.

5 The 1988 scales were as a product of the Battlefield Attrition Study, (BAS) and the Review of Ammunition Rates and Scales (RARS). The reasonings are recorded on page 417 of the UK Staff Officers Handbook 1988, Army Code 71038.

6 In terms of cost, it should be noted that the Russian Zala Lancet-3 claims only to cost $35,000.

7 Musgrave, John, *Firepower: Making 21st Century Warfare Decisive,* 2020. This work is an outstandingly useful examination of modern warfare overall. It should be required reading.

8

AIR DEFENCE AND AIR SUPPORT

Air Defence and Air Support are considered together as they form two sides of the same coin. One defends against what the other employs. As has already been emphasised, freedom of action in the air is critical to modern operations, as it's denial to the enemy.

Air Defence has risen substantially in prominence since its near demise post-Cold War, mostly in response to the perceived need for the "drone threat." Much of this has been substantially overstated. As previously discussed, UAS that can carry capable payloads are usually 50-55kg. These are detectable by electro-optical systems, and radar optimised to search for UAS. They can be engaged by MANPAD systems such as Starstreak, Martlet or SA-7 types. Russian Orlan-10s are only 15kg systems and have been successfully targeted by such systems.

The primary threat to ground forces is not UAS but manned aircraft capable of delivering 1-8,000kg of disposable ordnance. Killing UAS is not unimportant, but killing their operators is preferable, so killing UAS is not just an air defence task. Using EW and Fires to locate and destroy UAS complexes and facilities on the ground should be the start point.

Small UAS, if detected, can be addressed by hand-held weapons using fire control optics such as SmartShooter®. The same applies to remote weapons stations with the same software and sensor capability.

A modern networked and dispersed air defence system aims to detect and kill enemy UAS, helicopters, fast jets and cruise missiles. Counter-rocket and Mortar systems (C_RAM) rose to prominence in Iraq and Afghanistan as gun systems that were able to target incoming mortar bombs and rockets, as the name implies. The flow-down result of this technology was systems like Skyranger 35. This uses a 35mm cannon mounted in a self-contained turret on an AFV or truck, able to engage anything from small UAS to fast jets within 4,000m using a combination of radar and electro-optics able to detect small UAS at 5km or fast jets and helicopters at 20km. Guns are notionally cost-efficient ways of killing any air threats short of fast jets. Gun systems attract more maintenance support compared to missiles. They also attract greater logistic debt in terms of the weight of ammunition. Flakpanzer Gepard fired at 1,100 rounds per minute from two 35mm cannon. This equates to 1.7 tonnes of ammunition a minute.

In terms of technology, enhancing any cannon-mounting remote weapon station or powered turret with a fire control system to have some anti-air capability is not hard. It just requires the right sensor integration with the fire control system. It is possible to give most IFVs this capability. The most obvious example, albeit with no fire control system, is the BMP-2, specifically designed to engage low-flying aircraft and helicopters with a 30mm cannon enabled by 1PZ-3 air defence sight for the commander. Post-1973 experience convinced the Soviet Army that a 30mm gave the IFV an anti-air capability but axiomatically made it more flexible than the anti-tank dedicated BMP-1.

Guns are short-range air defence systems. They are most likely platform-based or towed systems. In sharp contrast, MANPADs, or man-portable air defence, are missile-based and descend from the US Redeye and the Soviet SAM-7 of the 1960s. In terms of modern capability, there are at least 7-10 major types to consider, but the differentiating factor is guidance, of which there are two basic types. The most common is infra-red, as in "heat seeking" common to all Russian, Chinese, and North Korean types in service today. This also includes the US Stinger and the French Mistral. The less common, more expensive, but more capable systems are RBS-70 and Starstreak/LMM, which both use laser-based systems, albeit each is slightly different. Regardless, each required the operator to track the target, compared to the IR-guided weapons, which are "fire and forget". RBS-70 and Starstreak outrange the IR-guided types both in reach and altitude.

Both laser-based systems are immune to current countermeasures but have a significant training burden for skilled operators. This has been mitigated to a degree with simulation. Still, the time and effort are significantly more than that which the Russians would give to an SA-16 operator or that which the SAS soldiers who downed an Argentine helicopter and ground attack aircraft in the Falklands had been given using a US FIM-92 Stinger.

The real point of discussion for MANPADS is what their name implies. They are platform agnostic, so they can appear anywhere on the battlefield and be easily concealed. Multiple firing posts, crews, and reloads can be carried on one light truck.

They can be carried to the top of mountains or tall buildings. That can even be operated off rafts and fishing boats close to shore. They can be mounted on almost any light vehicle with multiple ready missiles and networked into an area air defence system.

A system such as RBS-70 could use as little as six firing posts to screen a 50km frontage against most air threats operating at low to medium altitudes.

The next step in Air Defence is surface-to-air missiles, or SAM systems, or GBAD—ground-based air defence. The technology here essentially consists of a radar, a missile, and a communications system. The number and diversity of the current Western SAM system prohibit any detailed examination, but several general points need to be considered.

The first is the radar, which is the backbone of the detection system and can be displaced a considerable distance from the missile firing units because of modern command links. Modern radars differ a great deal from those of even the 1980s. With modern systems, the radar might be completely agnostic of the missile it is cueing to fire.

The radar can provide targeting data and guidance if the missile is on the network. The key point about modern 3D actively scanned electronic array radars (AESA) is they are far more than air defence assets. They can detect enemy artillery by identifying the firing points of enemy rocket launchers and guns, and one radar can provide surveillance and fire control. The radars inherent to modern air defence systems have become significantly more flexible, thus more useful than limited to air defence. This makes them a vital component of any Fires Lead Command system because they are continuously feeding targeting data on which decisions can be based. The Israeli-made ELM-2084 reportedly costs about $15m per radar. It can look out to 470km and track over 1,000 individual targets. It can be transported on a 4-tonne truck or ground mounted. If the $15m figure is accurate, radar costs less than a Main Battle Tank. Suppose a division was scaled with ten such radars networked into the command system with the appropriate communications system. In that case, this begins to lean towards being a decisive capability. While radars should be easy to detect with electronic support measures, targeting a radar that may be 1-200km away in a way that the same or other radars will not detect and thus switch off and move is far less easy.

Gaining the true potential of such systems means having the skill, command, and communications networks to make them work, which requires considerable training to obtain and maintain.

To revert to the air defence discussion, what digitised command combined with AESA radars now offers is air defenders monitoring an all-informed command picture which can filter out every other feed except air threats, which the air defenders can cue the appropriate missiles to engage. That missile could be several different types, all existing within the same network, gun-based systems, and MANPADS. Given the ability to generate such all-informed target pictures, it is a small step to leverage the same information to generate air deconfliction for your own UAS, helicopters and medium to low-level fast jets from artillery and rocket fires, thus speeding up the divisional deep and close air support and fire missions. This is only if you deem artillery deconfliction necessary instead of taking risks to enhance performance.

As stated several times thus far, air and spectrum freedom of action is essential to modern operations because exploiting that freedom of action for the wider benefits of target detection and communications inherent to it allows commanders to get the most from limited resources. Gaining freedom of action and making the best use of freedom of action are not related. Each is a distinct and separate skill.

The critical point to understand about the air defence organisation in this discussion is that we are only concerned with providing air defence for the Division. This limits the equipment set to short-range air defence systems. The inclusion of longer-range systems complicates command, control, and deconfliction and adds cost, weight, and complexity.

Theatre or Corps-level air defence responsibility will lie well outside the division's range of tasks.

The Air Defence Unit

Again, the air defence unit is broadly like all battery-based units. The notable difference is that each air defence battery comprises 1 HQ/radar vehicle, six gun-based systems such as the Skyrangers seen in the Cavalry unit, and six vehicle-mounted SACLOS or Beam-riding MANPADs with a dismount capability. The HQ/Radar vehicle uses a lightweight 3D tactical air defence radar with a 100km, allowing for UAS detection at 20km and fast jet at 40km. There are also 4 EPLS racks of re-supply per battery.

The Air Defence unit works closely with the ISTAR Unit regarding what radars are switched on, when and why, to prevent them from being detected or attacked. For that reason, the bias within the engagement envelope of the air defence weapons themselves is towards electro-optic targeting. The other key reason for close cooperation is to prevent the accidental targeting of the ISTAR unit's own UAS systems.

Air Support

As stated many times, unless you have freedom of air action or are at least able to exploit it when it occurs, most land forces will lose rapidly and decisively. Thus, the major emphasis of all Air Force activity is gaining the freedom of action to support ground forces, so the objectives of the Air Forces should be fused with those of the Land Forces.

Counter-air operations are explicitly designed to allow for Air Forces to conduct ground attack, close air support and interdiction. Close air support and even battlefield interdiction are primarily command and control problems. Ground commanders need aircraft to strike the right targets at the right time, ideally with the right weapon or the weapon the air force deems appropriate. Aircraft and air-delivered weapons are a very limited resource. All fast jets must be able to perform all missions and deliver all weapons, so almost every successful fighter since 1916 has developed into or spawned a variant capable of delivering air-to-ground weapons. Interdiction or close air support, ground attack has evolved into a method for delivering precision weapons. Hence, aircraft rely on targeting data derived from their sensors, UAS and/or ground callsigns.

This expands the utility of close air support into a task not limited by weather or light conditions or having the aircraft within the visual range of the target. Aircraft are not an alternative or substitute for artillery, nor should anyone aspire for that to be the case. Unlike WW2, Korea or Vietnam, modern aircraft, once lost, are unlikely to be replaced within any useful period. The Air Force's mission will and should remain to gain air freedom of movement and destroy enemy air defences to enable the destruction and degradation of enemy ground forces, command, infrastructure and logistics to ensure their operations are severely impeded. The potential enhancement of air forces regarding campaign relevance is that sophisticated long-range stand-off weapons do not necessarily require high performance, thus expensive delivery platforms, as they are operating outside the enemy's air defence threat envelope.

There are, of course, other critical air missions we will discuss later.

Attack and Armed helicopters exist as Air Force and Army assets, depending on which nation is being discussed. In Israel, the AH-64 Apache is an Air Force asset; in the UK, they belong to the Army. Helicopters are and always have been inherently vulnerable. Their survival is directly related to low exposure, night operation, and stand-off targeting. They are primarily an anti-tank platform that has now morphed into a system optimised for delivering precision weapons, much like fast jets. You do not need a $40-million attack helicopter to deliver a stand-off precision attack. Simpler, cheaper aircraft will suffice for lower running costs and more austere support. Modern night vision and electro-optics give that almost any helicopter can be upgraded to fly under the same light and weather conditions that attack types can. The cost, weight and complexity triangle is highly relevant. The missiles coming off the rails will be the same regardless of airframe type. Clearances and qualifications permitting, an EC-145M or MD-530 can deliver the same long-range standoff ATGM as AH-64 or other dedicated attack types.

Depending on context, armed or attack helicopters are best considered an anti-armour reserve or precision attack platform. The idea of any armed helicopter attempting close air support rolling in with rockets and guns seems only an idea persisting in extremis or an enemy with no air defence capability whatsoever.

Support helicopters are subject to a completely different set of arguments. Currently, the UH-60 and the CH-47 are still the dominant types in most Western nations.

These are 50-and 60-year-old designs, respectively. Support helicopters are an essential part of any modern land force, even if it is just a small number of single-engine types. Helicopters are almost non-discretionary to modern operations, particularly regarding medical evacuation. Still, many nations have flirted with or committed to the "air manoeuvre" idea, which will also be discussed later.

Support helicopters are extremely high-value assets. This does not prohibit their employment in high-risk environments, but it does mean that their employment needs to be as highly consequential as their value.

9

ENGINEERS

For this part of the discussion, we will mostly focus on mobility and counter mobility, but with some consideration of fortifications and infrastructure. This will consist of:

- Essential Offensive tasks
  - Wide wet gap crossing
  - Find and breach obstacle belts
- Essential Defensive tasks
  - Dig 3m x 3m ditch and berm
  - Lay Minefields
  - Dig Fortifications
  - Build Infrastructure.

River and Gap Crossing

The river crossing was probably the skill that created military engineering as a distinct arm or service, but it is important to distinguish the technical capabilities found in combat engineering units as something distinct from river or "wide wet gap crossings." The distinction between rivers and wide wet gap crossing is useful to the degree that vehicle-laid bridges, whether armoured or truck-launched, will get across many small rivers of less than 20m. That will also cover a lot of drainage ditches, canals, small wadis, and washouts. That will also cover the gaps left by bridges destroyed by demolitions. The capacity to manoeuvre is inextricably linked to that of obstacle crossing. To some extent, much about combat engineering subverts the cost, weight and complexity framework advocated thus far. An anti-tank ditch can be addressed quickly by merely bulldozing spoil into it. The same is true of drainage ditches, albeit the blocking of any flow could well result in flooding, which would probably become more of a problem than the ditch itself. The British system of plastic fascine pipes does have considerable merit in that regard but requires large armoured vehicles to manoeuvre in the sort of bundles that can usefully address a 2m x 2m ditch.

A 20m wide and 3m deep river stops every military formation, bar those with amphibious vehicles. This was so critical to Soviet and Russian military thinking that almost all BTR and BMP APCs, IFVs, and some SAM systems were intended to be amphibious.

An armoured vehicle-launched bridge (AVLB), such as the Leopard 2-based Leguan, can lay a 14-or 28-meter bridge with an 80-tonne capacity. The Finnish Army uses Leguan from a Sisu 10x10 truck. A lighter truck-mounted system, such as REBS, a rapidly emplaced bridging system, can lay a 13-meter bridge with a 50-tonne capacity. The French Army has a specialised 10x10 truck to launch a 28-meter 85-tonne capacity bridge called SPRAT.

General Dynamics has also mounted the 22m Beaver bridge on 8x8 and 10x10 MX2 truck chassis. There is no shortage of truck-based gap-crossing bridges at under 40 tonnes GVW.

There is an argument for permanently attaching AVLBs to the tank or AFV units on which their chassis is based. This is a logical variation of the permanently attached infantry in combined MBT and IFV units. The British did trial this idea, at least for a short while, but according to one Royal Engineer officer, the concept failed as the Cavalry and Tank Regiments tended to give the training and command of the attached AVLBs a very low priority. Thus, the idea floundered.

Regarding long-span AVLBs, it can take up to 30 minutes for a subunit of 28 tracked vehicles to cross when accounting for tactical spacing and even more at night, with no lights and possibly bad weather. This means getting a battle group of 135 or more vehicles across the 20m wide river would require at least four bridges to cross within 30 minutes.

Therefore, a tracked vehicle armoured battle group can surmount the 20m river while accepting some delay. A light force unit would have considerably more difficulty without access to vehicle-launched bridging.

The most likely solution for light forces used to be the UK's Air Portable Ferry bridge, which used a combination of pontoons and decks stowed on four 18-tonne truck racks to make a 35-tonne ferry. This is no longer in service. The ferry took 12-13 men two hours to construct, but for a short gap, it would be a de facto bridge, reaching each bank.

That same system can suffice for "wide wet gaps" in rivers and lakes. This is an entirely different scale of problem, which necessitates specialist units. These generally break down into two distinct forms of equipment: ribbon bridge systems and amphibious rigs. Both can be configured as bridges or ferry systems.

Ribbon bridge systems are mostly variants of the Soviet PMM, and the direct Western equivalent is "IRB" or Improved Ribbon Bridge. The German M3 and its equivalents best exemplify amphibious rigs, such as the Turkish FNSS Otter (Samur).

The comparison between ribbon bridge and amphibious rig conforms to the Cost, Weight, and Complexity argument.

Ribbon bridges are cheaper than amphibious rigs but are more manpower-intensive and slower to launch and configure. They require specialist small boats to assemble and manoeuvre. A full bridge unit comprises pontoons, end sections and workboats. Plus, every element is delivered by an 18-tonne vehicle, which would normally have a common chassis with the main type of logistic truck, must move to a hide with its drivers, who take no more part in the operation. This all equates to a large amount of traffic on and off the bridging site before crossing operations begin. The pontoons themselves are incredibly resilient to fragmentation and damage and, thanks to a foam filling, stay afloat unless physically disassembled by a direct hit from bombs or artillery shells.

Amphibious rigs are expensive and maintenance intensive but require substantially less manpower and, in theory, can assemble a 100m bridge in less than 10 minutes. Given the same conditions, a ribbon bridge can reportedly take 30 minutes. The other amphibious rig advantage is that every vehicle is an identical unit, so it makes better use of platform numbers and enables the system to have some redundancy. Additionally, an amphibious rig bridge can split into ferries or rafts and vice versa far faster than is possible for IRB.

Most bridge crossing operations will transition from a raft or ferry-based assault phase to a continuous bridge phase to increase traffic rates and cut the crossing operation's exposure time. As is obvious, bridges can be more easily targeted than ferries or rafts, so the time taken to transition from ferry to bridge and possibly back again is not a trivial consideration. An additional consideration is that pontoon-based ferry and raft arrays can usually carry more vehicles than amphibious rigs.

High traffic rates associated with bridging sites may also require heavy vehicle matting to prevent the banks or approaches from becoming boggy and unpassable.

However, this may not be necessary if cut logs are substituted.

While not a common military skill, a field engineer's ability to fell trees and process timber to produce wood for fortifications or structures should be considered part of an engineer's overall capability, especially in Europe or heavily forested areas.

Mines

Minefields are likewise inherent to manoeuvre. Minefields are cheap and easy to lay, but there has been a considerable erosion of expertise and resources allocated to mine warfare since the end of the Cold War. What follows will focus on anti-vehicle mines and not anti-personnel mines, not because of any legal or moral qualms against their employment but because anti-personnel mines are only effective under certain conditions and will be discussed in the later chapter dealing with those conditions. My experience having trained with the C3 Elsie anti-personnel mines during my time in the British Army was that their employment had not been adequately considered. Thus, excluding certain special circumstances, they might have presented more of a risk to our forces than to the enemy. That is not to say they weren't potentially very useful, but considerable care was needed in terms where, when and how to use them, if removed from the context of the end of the world, "World War Three!"

Any army serious about land warfare must have solid expertise and confidence in employing anti-tank mines. Not adequately accounted for in the history of WW2 and only recently relearned in Ukraine, buried anti-tank mines tend to be disproportionately effective in terms of reducing an enemy's freedom of action. The problem is that they no longer exist in most Western Armies. The only published exceptions are Israel and the US. Russian, China, Iran and North Korea almost certainly have millions stockpiled, and Russia has laid a considerable quantity in the last two years.

While no longer in service, the potential of the British L9 Barmine system is worthy of deep understanding. 1.2 meters in length and 11 to 8 centimetres square, the L9 contained 8.4 kilograms of explosives for a total weight of 11 kilograms. At the time of introduction, it would disable every main battle tank or AFV in service. No open-source experience suggests it would not immobilise most modern AFVs, even those intended to be mine-protected. The body of the mine was a plastic extrusion making it extremely hard to detect. It could be laid underwater to a depth of 1m. The bar configuration of the mine made it substantially more efficient than circular mines in terms of the chance of hit being initiated and, more importantly, storage and carried weight. One NATO standard pallet carried 73 barmines, which equated to 360m of one row of a high-density minefield.

The current US Army effort is towards "smart mines," as in the XM204, a 60cm x 60cm 35kg box capable of projecting up to four sensor-fused top attack submunitions to a detection radius of 100m, using Doppler radar for targeting. There is little point in comparison in terms of cost, weight and complexity. The Russians have developed a similar, far smaller, simpler "smart mine" with the PTKM-1R. As of December 2022, it has seen some employment in Ukraine.

There are also scatterable mines, be they artillery, rocket, or helicopter-spread mines. Artillery or rocket-launched systems could arguably be said to be part of an overall fires capability, so they are not part of a counter-mobility, albeit good Command and staff should be able to resolve that.

Regardless of type, all systems have an underlying dynamic:

1. Cost and weight
2. Time and effort are needed to emplace
3. The overall effect of retardation or fixing.

While obvious, cost and weight need to be considered from the standpoint of limited resources. It is entirely possible to lay minefields that never inflict casualties. Only 1-3 percent of mines in a barrier minefield belt might cause casualties. Thus, the argument for smart mines becomes more relevant.

Time and effort needed to emplace is a major consideration. APC-drawn mine ploughs could lay Barmines at a rate of about 8-10 per minute, but hand laying was needed for effective concealment. Smart mines can be rapidly emplaced at a far lower density, requiring less time and manpower. They can also be uplifted easily if not used, meaning they get recovered, which is extremely time-consuming with buried mines. Still, the ability to recover and regenerate laid minefields is considered the mark of a skilled Army. However, this only applies if the enemy does not take possession of them first or destroy them as part of EOD clearance. Scatterable mines can be placed quickly and ideally in front, within or behind an already-identified moving enemy. Most modern scatterable mines have inbuilt self-destruct systems which trigger or render them inert after a given time.

The decisive effect sought is that of retardation (slowing) or fixing the enemy so that they can be struck by other means. All obstacles are covered by fire, so the real effect of a minefield is to slow or halt the enemy so that fires can gain the greatest effect and inflict the greatest number of casualties.

Recent modelling by the British Army showed,

- Tempo: pace of close combat reduced by **60 percent**
- Rate of advance reduced by up to **60 percent**
- Covering fire is **twice** as effective
- Systems of obstacle types work best in combination (ditching, plus mines, plus wire etc).

Overall, the effects varied widely in terms of casualties inflicted directly by mines, but the presence of mines seems nearly always decisive. [1]

Breaching

Breaching minefields is an issue of speed and protection regarding methods and equipment. Two systems of breaching commonly exist:Mechanical – Rollers, ploughs, flails and bladesExplosive – Fuel Air Explosive and Explosive hose.

Mechanical systems are best employed either by dedicated combat engineer vehicles or MBTs adapted to use rollers and ploughs. Rollers and ploughs must be carried on trucks until required and then need time to fit. Each system has its own merits, depending on the terrain. Ploughs will work in sand and soil but will not cope with rocky terrain, and ploughs need extremely powerful vehicles using large amounts of engine power. Rollers can be used on almost any AFV but are vulnerable to double-impulse mine fuses. Deep full-width plough systems also required a good deal of horsepower and torque work. Blades can usually cope with surface-laid mines, but most MBT-mounted blade systems are for earthmoving, not mine clearance. The exception is armoured military D9 bulldozers, which can clear 4.3m wide lanes to 30-45cm depth. Beyond breaching, D9s can clear or create almost any obstacle, including defensive berms and tank firing ramps. It is often quicker to push dirt up than to dig down.

Current variants of Israeli D9s are capable of robotic operation, thus reducing personnel exposure. Almost any AFV, such as the British Terrier Armoured Engineer Vehicle, can now be adapted to operate unmanned and controlled remotely. The ideal may be tracked UGVs, but these would have to have the automotive power to conduct mechanical breaching.

The problem with mechanical breaching is that it is slow and exposes the vehicles and crews to both direct and indirect fires as well as any smart mines present. This is where a combination of mine types seriously compounds the task's difficulty.

Fuel Air Explosives (FAE) and explosive hoses simply destroy any type of mine within the affected area. The Israeli Carpet system uses large calibre rockets fired into a lane configuration using computer control. The rockets can also be employed against fortifications and large structures but from no further than about 160m. The system is normally mounted on a heavy armoured chassis as a Puma or Namer, but it can also be used from a trailer towed system. Explosive hose systems also conform to the same basic configurations, the most notable being the US Army's M1150 ABV based on the M1 Abrams chassis, which can use two explosive hose sets and mount a full-width deep mine plough.

Obstacles

Obstacle creation is another overlooked aspect of modern warfare. You don't need to demolish a bridge if you can dig out the approach ramps so that no vehicle can reach them. A tank ditch of 3m width by 3m deep is a significant obstacle requiring specialist engineering equipment, such as an AVLB or heavy earth-moving equipment, to cross.

As a baseline requirement, an excavator should be able to dig 40-100m of 3m x 3m anti-ditch in 8 hours.[2] 8m per hour equates to 96m in 12 hours but is highly dependent on soil type.

Obstacle creation is not so much an equipment capability as something born of experience and deep understanding but requires earth-moving equipment and careful planning. As with minefield, the intention is rarely, if ever, to block the enemy's progress but more to slow him to where he can be targeted. There are few routes short of multilane concrete motorways or autobahns that an excavator cannot deny in 20 minutes.

Excavation and explosive demolition capability will account for almost any anti-tank, thus AFV obstacle a commander wishes to create. As previously emphasised, at least in defence, the obstacle and fire plan aims are identical. Both should be the product of one mind.

The ability to rapidly dig entrenchments and below-surface vehicle slots cannot be dismissed as a requirement. A 2m x 2m x 7m vehicle slot takes just under 30 minutes to dig, so four excavators will dig 4 slots per hour. This means a whole unit or 100 vehicles can be dug in 24 hours of continuous work, allowing for the spare manpower capacity to work 12-hour shifts.

Many other land force engineering capabilities need consideration, from creating hard standings for helicopter operations and field hospitals

to restoring civilian power and water supplies and building refugee and
POW camps. However, all those capabilities do not necessarily require
military personnel, so the investment focus should fall towards engineering
tasks that support breaking the enemy's will.

The Field Engineer Unit

The Field Engineer unit aims to generate four sub-units for brigades or
units as required. A sub-unit contains two truck-mounted excavators., two
truck-mounted bridges, and two D9 Bulldozers on 77-tonne low loaders.
There are also 12x12 14-tonne UGVs optimised for breaching operations
equipped with active protection systems. These will be carried on EPLS
racks mounted on 8x8 trucks long with the operators. Each sub-unit also
has 3 EPLS racks of engineer stores available, such as mines, defence stores
and specialist tools as required. Lastly, there are two armoured personnel
carriers, each with eight trained engineers to assist as required.

The River Crossing Unit

The river crossing unit uses the Improved Ribbon Bridge system to build
five rafts, each comprising one end section, four pontoon bays, and two
work boats. The rafts are assembled and operated by 16 engineers each.
The unit can also assemble a ribbon bridge to cross a 140-meter-wide river.
The River Crossing Unit also holds 20 Assault boats with small outboard
motors or paddles able to move 6-8 men, depending on the carried loads.

Endnotes

1 Personal correspondence with Major Mark Davies RE, - author of Countering Enemy
Mobility Redux, *British Army Review*, Summer 2023.
2 This number assumes two digging cycles per minute with a bucket volume 0.6m3.

10

COMBAT SERVICE SUPPORT

Combat service support has traditionally attracted little attention in wider military thought. Given its criticality, this is somewhat surprising, but given that most military thoughts emanated from men from combat arms, maybe it should not be. The old and oft-repeated adage that "amateurs talk tactics and professionals talk logistics" is no less true than it is harsh and should be given the appropriate context of the fact that General Archibald Wavell's patient reminder to Basil Liddell-Hart that military historians mostly neglect the effects of fear, hunger, weather, and lack of sleep. It has been my direct experience, supported by historical study, that a strong indicator of a good army is combat service support and how well command and staff account for it. My force development modelling supporting two British Army projects showed that logistic support regarding the number of trucks and attendant planning considerations had impacts that even experienced senior officers who had joined the Army after the Cold War found surprising. It is extremely important to understand that every aspect of Cold War logistics, thus service support, was predicated on an existential war for which there was no plan for tomorrow. This needs to be contrasted with the Israeli view apparent in the 1960s and 70s that existential wars may recur year after year.

Medical

Everybody knows medical evacuation and support are important, but that can leave many assumptions unexamined. As previously explained, the dead have a disproportionate effect on the political will for a government to sustain an Army in conflict compared to just the wounded. The death of one soldier will be reported and commented on. In sharp contrast, the seriously injured may never attract comment. Body armour has previously been discussed as have the political implications it attracts. It is entirely right to observe that the wounded are also a far greater burden than the dead to any medical and logistic system. While a front-line soldier may need an allocation of 6-10 litres of water a day, as a casualty, that demand can triple or more in terms of the overall consumption of water needed by any field medical facility. The laundry demands of medical facilities are likewise often underestimated until experience kicks in, as is the need for cold storage and the maintenance of intricate medical equipment. Before any casualties are incurred, the medical elements of any grouping need a dedicated logistics pipeline.

The other issue of military medicine, often neglected but strongly implied by the political significance of the dead, is that casualty aversion and care have political ramifications. This has two expressions: the levels of force protection allocated to the force and the second is medical timeline planning.

No armed force wants to suffer needless casualties, but casualty avoidance has cost, weight, and training implications. For example, giving all your logistics and deployed support vehicles some level of armour protection has a substantial cost, as does constructing hardened accommodations and facilities instead of tents. This can be a straight military judgement and deemed best practice, but all aspects have cost weight and complexity implications. A tented camp which might reach an initial operating capability of a couple of days could well be weeks for protected or hardened accommodation plus the diversion of significant manpower and resources. These are not trivial issues; both have had current and historical impacts on the British Army and Israeli Defence Force.

Medical timelines are likewise context-dependent, as in the time between a soldier being wounded and reaching a medical facility, which can stabilise the casualty for life-saving surgery and critical care. The commonly accepted time is one hour. Depending on the level of trauma care available due to trained medical personnel available close to the point of wounding or at the casualty exchange point, that might be extended due to military necessity. Still, there are wider implications for the overall conduct of military operations. Medical timelines extend the same constraints so that a force cannot disperse or exploit beyond its communications, fire support and logistics. We have previously looked at medical support at the unit level, but evacuating from the unit to a medical facility demands various approaches, some of which may be untested.

The US involvement in Vietnam confirmed the utility of helicopter evacuation, and it has since become the norm for almost every subsequent conflict involving Western Armies.

In recent years, the norm for both the UK and Israel has been helicopter evacuation from near the point of wounding to a fixed medical facility with surgical capability or, in the case of Israel, civilian hospitals. How viable air evacuation will be in the future is impossible to predict, as is the type of conflict, so the training and resource requirements should be biased towards no air evacuation. There is a need to stabilise as far forward as possible since even ground evacuation may be hazardous or even impossible, given any enemy capability to conduct a deep battle similar to that we envisage.

As previously discussed, the most likely utility of UGVs is casualty evacuation to the rear, having brought ammunition forward. Casualty evacuation from the point of wounding can rapidly deplete a platoon's combat power, so a vehicle that needs no more than one person to operate a project will have undoubted merit. Still, it seems likely that under most circumstances, casualties will be evacuated using the armoured vehicle that brought them forward.

Most casualties do not require immediate lifesaving intervention, thus the triage system. Depending on the existing policy, those that do generally fall into three groups:

P/T-1 Requires immediate surgical intervention and needs immediate evacuation

P/T-2 Need intervention within 2 hours – medical timeline dependant

P/T-3 Requires evacuation when possible.

The faster the casualty can be moved to a secure location where a dedicated surgical intervention can occur, the better. Assuming catastrophic bleeding has been stopped and the casualty is still breathing, on reaching a secure location, the immediate requirement will be to stop any remaining bleeding, support the airway, and maintain respiration. Then, supply replacement whole blood and any additional pain management.

There is the possibility of bringing medical care forward when evacuation is likely interdicted or requires excessive time. In terms of vehicles, this means armoured ambulances with onboard oxygen, monitoring equipment, and the ability to store whole blood. An APC with a couple of stretchers in the back will no longer suffice, albeit these will still be needed to evacuate T-3 casualties. As a planning assumption there seems to be merit in aiming to support each battlegroup in contact with a detachment able to stop bleeding, support the airway and maintain respiration.

Under the Laws of Armed Conflict to which all Democracies subscribe, there is an absolute requirement to treat enemy prisoners of war or detainees with the same level of care afforded to their own troops. So, while the attack may succeed with few of your casualties, many enemy casualties will need treatment.

The aspiration should be to avoid large, hard-to-move field hospitals for warfighting operations. However, Field Hospitals can have substantial political merit because they can be deployed to natural disasters or refugee crises. They also allow militaries to provide non-lethal military support to international efforts when sending combat forces would be politically contentious.

A Field Hospital potentially fixes the Division, so it would be all too easy to focus on evacuation, thus delegating the problem to the Corps or higher. The problem here is circumstances such as the Falklands War where off-loading the problem was impossible as one Field Hospital had to support two infantry brigades conducting dismounted close combat. If medical timelines are the policy to be supported, then mobile surgical facilities operating as far forward as possible seem to be a non-discretionary requirement.

"Role 2" Field Hospitals enhanced to conduct surgery under general anaesthetic and provide limited intensive care before evacuation to a larger facility (Role 3) are an option to consider. Regarding warfighting expeditionary operations, the ambition should be evacuation out of the theatre as soon as possible. For national defence, the aim should be to get the casualties to civilian hospitals as soon as possible. Care beyond that falls outside the scope of this work.

Role 2 Enhanced Hospitals can be designed around a description of capacity in terms of the number of operating theatres, intensive care beds, and beds on a ward. A 2,4,20 would have two operating theatres, four intensive care beds, and 20 beds for sickness not requiring evacuation. The time it is safe to move someone after surgery varies considerably, but evacuating even intensive care cases from theatre should receive national priority. It would thus fall to the level of command above the Division to clear casualties as quickly as possible.

Field Hospitals require extensive training in deploying, building, managing, breaking down, and recovering, especially in dynamic and uncertain situations. If an expeditionary capability is required, tented or containerised facilities must withstand almost any weather. The fuel demands for heating and air conditioning are not trivial, so vehicle-based or trailer generators may be required.

Such facilities operate 24 hours a day, so they need sleeping, laundry, catering, and washing facilities with appropriate medical hygiene standards.

Additional thought must also be given to unit-wide outbreaks of vomiting and diarrhoea, a common occurrence in even modern deployed armies, and infectious diseases requiring quarantine measures. The medical capabilities beyond this needed for the general health of the deployed force increase the size of the medical facility and its staff. While dental treatments are important, gynaecology and paediatrics are examples of facilities generally only relevant to civil-military affairs and disaster relief.

While not strictly medical, a division or similar grouping needs to have the capacity to rest and restore personnel who may have spent 14-30 days in combat and harsh field conditions. This means showering, laundry, re-issue of equipment, fresh hot meals, and uninterrupted sleep in clean, comfortable, and safe conditions.

For example, mobile shower and laundry units are best understood as something prioritised by Field Hospitals, but planned spare capacity can and should be used to support units recovering and regenerating away from the front line.

The Medical Unit

The Medical unit would be primarily organised around three sub-units: Surgical, Support, and Evacuation Companies. In addition to these, the normal command and service support entities would be present, including an HQ facility with robust communications support to ensure records, requests, and demands are fulfilled.

The Surgical company would have 12 demountable expandable container-based vehicles to sustain two operating theatres and four intensive care beds. This accounts for four container systems, with the others providing two with ten general care beds, an X-ray facility, a laboratory, a scrub and washing facility, temperature-controlled storage, and power generation. Designing hospitals is a dark art for which specialist civilian companies exist, so this is notional for discussing purposes.

The support company provides accommodation for medical staff, catering for everyone, laundry, and shower facilities, all based in 18 containers with some attendant tentage.

There would also have to be some mortuary affairs capacity which would focus on identification, limited investigation into the cause of death, and temporary interment in case evacuation is not possible.

The evacuation company has 32 vehicles, split into eight four-vehicle detachments using the wheel APC from the infantry. In each detachment, one vehicle is equipped with critical care equipment such as oxygen and respirators to evacuate two critical casualties, and three are each equipped to evacuate a combination of four stretchers or six walking wounded. Each vehicle is crewed by three staff, at least two of which can provide wound non-surgical care in cleaning, closure and dressing. Detachments would focus their effort towards those units in contact, so semi-permanent attachments are not the intention. In extremis, it may also be the case that the detachments may well be clearing the field hospital by moving casualties to the beach, seaport, railhead or airhead.

The overall evacuation capacity is 16 P1/2 with 192 stretcher cases.

Logistics

Logistics can rightly claim to be the decisive factor in most campaigns, all else being equal. Therefore, logistics needs highly skilled and determined commanders and staff officers, or all else is for nought. Logistics as a performance point can be reduced to the ability to move mass over distance over time. Thus, if it takes 10 hours to move 1,000 tonnes 100 kilometres, the delivery rate is 1 tonne per hour. Five thousand tonnes over 150km in 10 hours is 3.3 tonnes per hour. Fairly simple arithmetic or a spreadsheet can tell you the required vehicle lifts based on accurate convoy speeds and loading and unloading estimates. This may seem like a "statement of the bleeding obvious" (SOBO), as should the realisation not to design units, formations and divisions or plan operations which you cannot sustain over time and distance. Yet both historically and today, command staff and force developers often fail to account for such basics.

Taking water and rations first, there will be a fixed number associated with each soldier in the unit in terms of what they consume in 24 hours. Climate is a factor, but why not plan for the worst case? Modern water filtration systems have altered water logistics and sustainment, where water is available and where filtration can provide potable water. A unit of 500 men will need at least 10 litres per man per 24 hours to account for drinking, centralised feeding and basic hygiene. That is 5,000 litres, and a unit should hold three days' worth of water, making the total hold 15,000 litres, which needs to be split between bulk water held in 1,000-litre pallet tanks at the unit level, and sub-unit trailers, water pallets and plastic bottled drinking water, which is usually 756 litres of water per pallet or 504 1.5 litre water bottles. Plastic water Jerry cans are robust and generally fail-safe. Still, from an administrative standpoint, it is often easier for soldiers to collect

4 x 1.5 plastic bottled water than to have a queue around the water trailer. The technological answer is that equipment transportable on one pallet can purify 5,000 litres a day from almost any water source and can easily be operated by unit or sub-unit support echelons.[1] The widespread issue of modern individual water filters should further alleviate the issue of water on logistic hold and handling, except in environments where water is scarce.

Excluding water and rations, the amount of sustainment a unit requires depends entirely on the type of force being supported, how it is grouped and organised, and the consumption rates for the required level of performance. For example, if the close support regiment has 18 guns intending to fire 350 rounds per gun daily, the requirement is to supply 6,300 rounds every 24 hours, equating to 37 8x8 truckloads.

The baseline requirement for most land forces is thus an 8x8 truck capable of moving EPLS (enhanced palletised loading system), a development of the UK's now obsolete DROPS racks (Demountable Off Load and Pick Up System). EPLS can carry a 20-foot ISO container or 10 NATO standard pallets with a payload capacity of just under 15 tonnes. This limits each pallet to being under 1.5 tonnes. It can also use a 6x6 trailer. A combination of racks, containers and pallets creates substantial flexibility. For example, the same truck can offload artillery ammunition and pick up a 10,000-litre fuel distribution rack. As war is the domain of chaos, flexibility matters. If the dedicated refuelling vehicle breaks down, what is the reversionary measure?

The NATO standard pallet is a deceptively simple item essential to modern logistics. The pallet measures 1m x 1.2m and can hold 1.8 tonnes, though, as previously noted, this should be kept to less than 1.5 tonnes. Still, it can also be a constraint once stores and ammunition are moved forward to the unit level. For example, a NATO standard pallet only holds 6 Javelin ATGW rounds. In contrast there are 9 TOW missiles on a pallet. Once removed from their pallet and stowage containers, many rounds can be held on light vehicles using specialised lockers and racks. So, instead of a truck with four pallets equating to 24 rounds, those 24 rounds can be easily loaded onto a light 4x4 pickup. Weight and volume often mean that units should avoid holding pallets on trucks. This seems counterintuitive until you realise that pallets require mechanical handling and that there is a subtle yet critical difference between holding and distribution. Generally speaking, a logistics system wants to reduce the number of physical interactions and transfers required from the depot to the user. There needs to be a balance between effectiveness and efficiency.

Once delivered to the unit, ammunition must be stripped off pallets and re-stowed to the distribution system that suits the unit's needs. In some cases, such as 120mm mortar ammunition remaining in the pallet maybe optimal. In other cases, such as M72 LAW (45 on a pallet), 7.62mm link or 40mm grenades, it will not. Two complete re-supplies of a rifle company's small arms ammunition can be held on a 2-tonne light truck, immediately available within the sub-unit.

Fuel is as critical as ammunition. How fuel is stored and distributed has a considerable impact on performance. Units and formations need to hold both bulk fuel and packed fuel. Packed fuel is fuel held in 20-litre cans or Jerry cans, a central distribution element. The logistic requirement for fuel is based on the Fuel Consumption Unit (FCU), which is the number of litres a fully-loaded vehicle consumes driving 100km on a hard, flat surface. Usually, no speed is specified, but a convoy speed of 40kph is often the underlying assumption. The reality of modern operations may be 80kph. Planners need to know how much fuel a unit needs to cover a certain distance, such as the number of litres needed to drive a convoy of 600km. There is also the amount of fuel needed to fill up a vehicle. This is called a "fill". Where the number of FCUs is less than "the fill", as will be the case for most tracked vehicles, extra fuel needs to be carried by the unit. Where the fill exceeds the required number of FCUs, the excess amount still must be held within the unit. The critical number is the total fuel a unit needs to fill the vehicles and the amount to be carried so that all vehicles can meet the required distance. That would usually be 4-800km. Units need to hold twice that figure in both bulk and as packed. Some readers will be familiar with NATO Staff planning guidance that specifies units should hold X amount of FCUs in bulk and X amount as packed. The problem is this often leaves a large amount of spare tank capacity not accounted for by vehicles that can move further than the convoy range demanded.

All the planning data must be tested realistically, and the actual performance limitations must be recorded.

Above the unit level, bulk fuel should be held on EPLS ISO rack systems or similar, either ground-dumped or on wheels, and they will need to be well concealed. A 210,000 US gallon bladder tank (794,936 litres) measures 22m x 22m. That can be detected from space and targeted. Smaller bladder systems should enable effective concealment and be inside ground-floor structures or underground. The issues associated with theatre fuel infrastructure fall outside the scope of this work but are critical to an army's overall performance.

Related to fuel consumption is spare parts consumption, which also requires consideration in terms of logistics. Any unit with armoured vehicles must hold a reserve of spare engines/power packs, track, and so on or any components that can be changed at the unit level. Staff data will generally define the quantity needed based on the mean time between failure (MTBF) or mean distance between failure (MDBF). A rule of thumb used in my work derived from conversations with Brigade equipment support officers is that 300 medium-weight tracked vehicles driving 400km will consume 18 EPLS racks of spares. This equates to 50 tracked vehicles needing five racks of spares to drive 600km.

This figure was derived from vehicles using metal tracks so band track would alter that figure considerably. What that number is is currently unknown, bringing us to the key point in logistics: understanding all logistics is about hard data based on rates of consumption and usage. Staff planning defines logistic requirements and staff planning data should be derived from real-world operations. Reducing wear on tracked vehicles is also relevant to maintenance considerations, so the aim to move tracked vehicles on transporters is of direct concern to the logistics unit. As discussed in Chapter Two, the IDF's Merkava IV tanks are transported on two-axle, 16-wheel trailers with a capacity of 77,500kg. This also allows for moving 18 NATO standard pallets, when not being used to transport tracked vehicles. This means that 111 such vehicles could move 2,000 pallet equivalents.

A key development in modern logistics is using digital data and reporting to reduce the number of transfers between the unit and the theatre point of supply. Ideally, any item arriving in the theatre should stay on the same vehicle or PLS rack until it reaches the unit that will use it. This is probably an unachievable ideal, but there is no reason why a truck from an infantry unit should not pull into a theatre-level supply depot and be directed to the racks already pre-loaded with the stores they require. Modern data tracking makes doing this trivially simple. Automatic data reporting can also enable a unit logistics holding to be tracked so that vehicles can be despatched with the exact required amount of resupply with machine learning automation tracking the variation in demand to provide a safety margin in case of a sudden increase in demand. The software, tracking and communications packages to enable this have existed for some time. The Rear HQ of the Division should have an hourly update on all resources held at the unit level once the various echelons have distributed the resources concerned.

The Logistic Unit

The Logistic Unit would comprise three sub-units of 20 vehicles plus the standard supporting detachments. There would be two Logistic units within the Division, meaning 120 logistic vehicles. If EPLS is the standard and equipped with trailers, these vehicles could lift and hold 2,400 pallets. A percentage of vehicles would need to carry 10,000-litre fuel racks.

Recovery and Repair

Vehicle and equipment maintenance and repair is a capability in and of itself. Most armies maintain maintenance detachments at the unit level. Light Aid Detachments in the British Army and "Petah" in the IDF. "Petah" because the Hebrew transliteration from "Fitter" relates to the IDF's origins within British training. Unit-level detachments will swap out engines and transmissions, repair armour packs, replace tracks, and fault-find electronic components to swap out the correct module or component. The limiting factor at the unit level is time and manpower. Suppose a vehicle cannot be fully operational within 8 hours. In that case, it might need to be backloaded to the maintenance unit, or the maintenance unit will bring up additional manpower, spares and tools forward if required because experience has shown that forward repair is preferable to backloading. Indeed, the IDF credited this approach as a war-winning capability in 1973.

One of the obvious ways to ease the maintenance burden is to procure simple equipment in terms of the cost, weight and complexity triad discussed thus far. Heavy-tracked vehicles will attract greater maintenance and repair burden than lighter-tracked or even wheeled vehicles. Automated gun-loading systems will fail more often than manual ones. That has to be balanced against the supposed reduction in crew numbers and rates of fire, if indeed either is important or decisive, which they may not be. As previously stated, what system or platform fails the most or attracts the most work will be a product of experience and will alter over time as vehicles age and wear out, so there is a strong correlation between repair, maintenance and cost of ownership, especially in regard to the procurement and storage of spare parts. As with all else, do not procure equipment you cannot afford to run and maintain.

One key data source can be Health and Usage Monitoring systems fitted to each vehicle. These systems use a range of digital sensors to record all aspects of vehicle performance. They can warn of impending failures, provided the failure has some detectable signature. While expensive, these systems can greatly reduce costs in the long run since they are a trackable source of reliable data.

Limiting the number of chassis held within the division and formation can also reduce cost. As of 1988, a British Army Armoured Infantry battalion only comprised six major vehicle types and major automotive components. With further thought and investment, that could have been reduced to five.

Maintenance units will also be responsible for recovery operations. Most vehicles will need nothing more than an 8x8 heavy recovery truck. Still, heavy armour will most likely require a dedicated detachment moving with the relevant units, equipped with vehicles such as the M88 or the UK's Challenger Recovery and Repair Vehicle (CRARV).

In the IDF tank crews focus on buddy recovery within the sub-unit, so at least righting an overturned vehicle or dragging a disabled vehicle to where it can be backloaded onto a transporter can be done without external assets.

Except for vehicles that have burned or had their hulls physically destroyed by gunfire or explosion, even small maintenance detachments can achieve something useful with enough lifting power, spares, tools, and technical knowledge. If components are swapped out within the unit, then the focus of the maintenance unit should be on repairing such components. There is a limit to this.

The Maintenance unit would mostly be concerned with recovering and repairing light weapons, communications devices, and sensors. Light weapon repair is simple and requires most component swaps. Communication devices and sensors will usually need to be returned to the manufacturer or depot, so the unit will despatch them and maintain a holding of spare units.

The Recovery and Repair Unit (RRU)

The RRU has three major sub-units:

- Vehicle LRU repair
- Recovery and distribution
- Technical Stores Holding and management.

The Vehicle LRU Company repair will repair power packs, weapon systems, vehicle sub-systems and sensors.

The Recovery and Distribution Company will help return vehicles to the unit, send forward repaired vehicles, and hold or maintain a small number of battle casualty replacement vehicles if available.

Technical Stores Holding and Management Company recovers broken weapons, communications, and sensor equipment and sends forward replacement items, which it maintains in stock.

Conclusion

Very few armies have ever been defeated with intact and functioning combat service support. It is entirely right to observe that combat units are wholly dependent on re-supply in every phase of war, and men who know they are being cared for and supported will achieve far more than those who do not. Combat service support is not just a sustainer of combat power but a very real creator of combat power.

Endnote

1 Karcher.

11

DEPLOY MARCH SUSTAIN

Readiness: Test of an Army

Readiness is the test of any army. If any army is not ready to move in 2 days or less, it is a force of doubtful political utility compared to one that can and is ready.

Readiness to fight inherently means having the ability to get to where the fighting needs to be done. That means, drive, fly, sail, or trains and maybe a combination of all four. Beyond anything else, all deployable land force elements know how to work with whatever method is employed. That means knowing how to load the unit onto trains, ships or aircraft. That means training to do it promptly, with little warning. Few other metrics compare in terms of what a good army looks like as it can move rapidly from its camp or base and either drive to the required area or embark onto aircraft, ship, and trains and then arrive in good order to immediately deploy to go to war. No more training can be done or is needed. You are either ready or not ready. Good armies are ready.

In the British Army, at the height of the Cold War, all units were out of their garrisons within 4 hours. This was achievable for a regular standing Army, most of which was garrisoned within 200km of its deployment area. For the Israeli Defence Force, in 1973, it meant a reservist getting from home to a tank turret in less than 24 hours. Most of the force was ready substantially quicker, albeit this was often achieved by less than elegant means. By the 1980s, how an Israeli unit deployed at speed had been substantially refined.

If you are an Army of National Defence, as with Israel, you need the capacity to deploy in less than 24 hours. As October 7th, 2023, showed, you may need some high-readiness units to respond in 1 hour or less.

Regarding October 7th, ten IDF brigades were immediately ready, with their reserve components joining them within 24 hours or less. Approximately ten more reserve brigades were ready for action within 72 hours. Approximately another 35 reserve brigades mobilised within 72 hours but required several weeks of refresher training to be operational.[1]

For this discussion, we will assume very high readiness is 2-5 days' notice to move, and high readiness is 6-14 days. If you cannot prepare your whole field force in 14 days, you are not serious about the task in hand. Moreover, any competent unit should always wait 2-5 days' notice to move. Moving all units from 2-5 days' notice to 12 hours for short periods should also be simple. What is the difference between 2 days and 12 hours?

Recalling officers and soldiers on leave or away on courses may be a constraint. Still, it is very unlikely that the absence of even a few key unit members will impact overall readiness. If it does, that indicates a training and organisation problem. Being at high readiness has the flow-down effect of being ready to train with little preparation because all your vehicles and equipment are at hand.

An Army should live at very high readiness. If you can't or don't, something is structurally wrong with your leadership, purpose, or funding. If your unit is not at high readiness, what does that mean about the overall approach to warfighting and fitness to task? If the claim is that it is too expensive, then at what point does the need to cut costs and expenses impact the overall training requirement?

At the unit level, readiness can be tested frequently and easily, albeit with some constraints. The most obvious is the issue of live ammunition, which is not a minor consideration. It can suffice that all the relevant sub-units report to the facilities concerned and go through the process with ballasted pallets or storage containers, which are retained for that purpose.

Living at high readiness has obvious implications, such as limiting who can go on leave and when and how quickly soldiers can return to the unit. In the days of mobile phones and near-universal communication, this is far less of a limitation than it traditionally was.

Sometimes, a unit must be downgraded from very high readiness to high readiness, but the aspiration should be to live at very high because, at that point, you are most useful to the government as you can be.

If there is a downside to high readiness beyond expense, then it is that it allows the government to delay decisions. Suppose the General

Staff warns the Government that it needs 30 days warning to deploy. In that case, the Government knows it needs to order the deployment well before the full implications of the emerging situation become clear. If the Government knows the Army can move with 2-5 days' notice, it will likely delay that decision until the last safe moment. Human nature attests to this.

Sea Deployment

Readiness and deployment are technically two separate things, but they are closely linked because readiness is the unit's preparedness to deploy. As stated before, it is no good driving to the docks to find your unit has never embarked on a ship before, and thus, all the friction and uncertainty inherent to something new and complicated come home to roost all at once. This means that the required action, be that embarkation onto ships or loading onto aircraft or trains, is familiar to enough soldiers in the unit to be explained and prepared for. This means this must have been rehearsed often enough that the relevant expertise is resident and current within the unit. If the Sub-unit commander last embarked his mechanised company on a ship as a Platoon Commander, then the relevant experience will be there. There is also a substantial difference between embarkation conducted as part of something planned months or years in advance with all the correct briefings, site and route reconnaissance and the same thing done at two days' notice.

The rapid deployment of armies rarely occurs under ideal conditions, so friction must be accounted for. The unit may get to the docks, but the ship is delayed for three days. So, the Commanding officer must think about how he might want to disperse his unit to make it less detectable to enemy satellites or international media and what training might be accomplished in the time available. Will sub-units use the field rations they carry, or will they feed centrally from company field kitchens, restaurants, or hotels?

Who needs to grant the authorisation to use the local economy? This may seem trivial, but those who have never faced such decisions will be less prepared than those who have.

As is obvious, deployment is primarily a command-and-staff problem that must be simplified to the degree that it is achievable even under the most challenging conditions.

Movement by sea should be conceptually simple, but much conspires to confound that ambition. In areas with robust vehicle ferry systems, such as the North Sea, Mediterranean, and the Southern Pacific archipelagos, vehicles and crews can drive onto a vessel, sail for less than 12 hours, and disembark easily. Nothing was that simple for the British Army sailing to the Falklands in 1982. Nor was deploying to the Gulf Wars.

For example, embarking all the vehicles for an Artillery unit is probably not complicated. Shipping ammunition and fuel is likewise simple. Delivering all these to secure a functioning container port with ample civilian contractors to store and position ammunition and with the vehicles ready for most personnel to be flown in and marry up with both is likewise simple staff and plans work. To deliver the same to barely functioning 300m of dockside in with insecure port area routinely under attack either by air or missiles combined with a refugee crisis is far more difficult. To deliver it over a beach is yet an additional level of complexity. Solving these problems is highly dependent on the type of shipping used. Thus, the real difference between a Land Force which can be moved from a secure port location to a secure port location, ideally with a secure airfield close by, has very little to do with the Land Force itself and far more to do with the investment in a shipping fleet that provides the capability to deploy globally or even regionally. As an ideal, the entire vehicle fleet must be able to drive, fully loaded, from the ship onto the shore. How this is done falls outside the scope of this work, but having a force that is physically capable of doing it is something that training, equipment and organisation can bridge.

Air Deployment

Deployment by air is another matter entirely but shares one common characteristic with sea deployment: the specifics of the available aircraft type are as important as the shipping types employed. For almost 50 years or more, the baseline assumption of air deployment was built around the C-130 for most nations. Additionally, the idea of the globally deployable force enabled by heavy transport aircraft has been debated from the end of WW2 until today. The US Stryker Brigades are a case in point. Aircraft have a constrained performance that dictates how far, how fast, and even how often they can fly. As a baseline data point, a cargo Boeing 777 can deliver 102 tonnes of stores 9,000km. Allowing for turnaround time, loading, and refuelling equate to flying >1,000 tonnes of stores or equipment between New York and Lagos in 14 days. This is one aircraft conducting ten flights. The sailing time between New York and Lagos is about 18 days. If anyone is curious why a cargo Boeing 777 is an example, an A400M can only fly

30 tonnes to 4.500km or 20 tonnes to 6,400km, and the Boeing C-17 can fly 71 tonnes to 4,480km. A C-130J can fly 18 tonnes to 3,800km.

All those aircraft have air-to-air refuelling but are considerably slower than a Boeing 777. A cargo Boeing 777 can carry 18 x 4x4 Land Rover-type vehicles, 5 M-113s or JLTV-type vehicles, and light trucks such as Unimog 435s. Configuring an effective unit around vehicles deployable by a wide-body, long-range cargo airliner is possible. Assuming an average base load of 100 vehicles per unit, then 20 sorties deploy that unit to the range of the aircraft, assuming a secure airhead with refuelling and cargo handling capacity.

The 2,000km March

In Chapter 2, we already discussed the equipment's capability to move 800km in 24 hours. Here, we must consider how that would be done and trained. 2,000km equals 60 hours of movement, rest, repair, and refuelling. To understand this problem set, we should assemble the Division in fields on each side of a major highway. Each unit is in an administrative laager so assembled in Company march columns. The aim would, therefore, be to march the Division to a similar site 2,000km away within 60 hours. Any formation can move 2,000km, but doing so in 60 hours is extremely demanding for anything other than small numbers of vehicles.

Many in the British Army will note that the 2,000km was originally a requirement of General Nick Carter's Strike Brigade concept, which I was initially deeply sceptical of until I did a fair amount of staff and planning analysis as to what it took to march a Brigade 2,000km at which point it became obvious that it was an extremely useful forcing mechanism for understanding logistics, sustainment and movement.

This is a test of performance, not of tactical acumen. If a division can do this, it is useful for many reasons, such as the skills and drills needed to sustain units on such a move and reach the other end fit to fight. The other critical point is that this problem is only presented in this way so we can discuss it and gain insights. As with the Strike Brigade, it was thus chosen as a performance requirement to better understand the Monash Division structure and choices. Do the planning estimates to see what your division or formation can march in a day. It will ruthlessly expose where the limitations in training, equipment and organisations.

A unit of 100 vehicles spaced at 10 per kilometre will have a pass time of 10 minutes if moving at 60 kph. Six units would pass in the hour, so the

Division of 20 units would take about 4 hours and 20 minutes. The total road length occupied by the Division would be 200km. This theoretical ideal assumes flat, good roads without traffic, turnings, intersections, bad weather, and nighttime. A convoy speed of 66kph or 41 mph should be viable for a vehicle with 100m of distance between it and the next and allow for a safe braking distance on dry roads in daylight with a fully loaded 8x8 truck.

Twenty-four hours of march would consist of 12 hours of movement, 4 hours of halts, feeding and maintenance, and 8 hours of sleep. Having all vehicles crewed by two or more drivers and equipping all vehicles with boiling vessels to allow for food preparation on the move, thus better use of time. Any overly constricting health and safety legislation should be dismissed, if possible, but not to the point where it endangers operational effectiveness. In the same way, it is negligent to overload soldiers; it is similarly negligent not to let them eat and sleep.

One evolution of this sequence is not a challenge. However, doing it for two and a half days will require pre-dumping the fuel availability along the route. Again, a 2,000km march may not be practical in all circumstances, but having a force that can do it is more capable overall than a force that cannot.

Signature and Sustainment

Convoys of large numbers of military vehicles can be detected via radar using moving target indicator (MTI) modes. This is nothing new as MTI is a 40-year-old technology, but active electronically scanned arrays have vastly reduced the size and weight of aircraft or UAS that can carry it. Any march formation must use very small packet sizes and 3-5 routes or more. The impact of such precautions must be balanced against the actual enemy's capability of possessing or even using such technology. That said, civilians with cell phones pose just as great a risk, or do they? Just because someone posts pictures on the internet of vehicles moving in one direction on one route does not reveal much if an effective deception plan is in place. The overall signature issue is inherent to the entire force, but it cannot rule over all else. Signature reduction is essential, but it can only be taken so far before it impacts the actual conduct of the operation. The ability to be of a size and configuration that can hide in most terrains or disperse to the degree that effective targeting is likely to be reduced must be considered. The current hype of the "transparent battlefield" or "no rear areas" should only be useful because awareness enables effective behaviours rather than paralysation. Activity must be balanced against security.

As previously stated, large high-signature "Fuel Farms" and logistic sites must become undetectable from any altitude and not visually ascertainable from as close as 100m. They need to be hidden. This is not hard to do. The real insight should be that whatever you are concerned about the enemy, he is more concerned about you.

The ability to conceal formations within terrains should be considered a priority regarding signature reduction and the ability to hide whenever at a halt. As mentioned, concerning training, that may even mean testing your force against your own sensors.

Formations' ability to remain undetected while static is a very high test of discipline and training. All the processes and procedures associated with high readiness must be conducted while avoiding detection. This may mean refuelling vehicles by hand, walking 5km for one hot meal daily, and then returning to your hide area. None of these things will be certain until they are done routinely and everyone concerned can be confident that this can be done safely and securely. Security will always be traded against activity, so signatures cannot be lowered to paralysis.

Endnote

1 Private Communication with Israeli Operational Analyst, approximated from open sources.

12

THE DEFENCE

The purpose of the defence is to preserve a condition or situation. The attack is to change it. It prevents the enemy from destroying your force or occupying a location. For all the nebulous sophistry about the attack and the defence, such as "the best form of defence is attack", you are still left with the absolute requirement to have either prevented the enemy from gaining a location or moving through an area, with your force still intact to the degree that it can continue with its task. Preserving the force should not usually equate to constant withdrawal, as the real objective of the defence is to defeat the enemy by breaking his will to continue offensive action.

The Defence in General

Any military operation is subject to almost infinite context; therefore, training must account for those actions and activities with the broadest and most widely relevant applications. These are:

- The Screen
- The Block
- The Counterattack.

The IDF uses a similar construct but is described as the security effort, holding effort and reserves.

This is common to most developed armies. There is little evidence that these have changed for over 100 years and are not likely to change. If, as an Army, you can successfully conduct these actions, you will probably be better prepared than others who cannot. The following is to discuss and debate what training will most likely deliver that result.

The desired course of events is for the screen to detect and then delay the enemy as they advance so that by the time they encounter the forces in the main defensive positions blocking their way, they have been set up for counterattacks and counterstrokes, which will defeat them. This is the ideal, not the real. Achieving this in training will lead to a generally more skilled and competent force able to adapt to the specific circumstances as required.

The overall Command or Divisional approach to defence should seize key terrain as soon as possible since reporting from a physical presence will allow the detailed planning that maps and aerial observation do not. It is equally important to generate and distribute orders based on battlespace to ensure that the relevant unit commanders clearly understand their tasks and missions if all else fails. Training for this means that all commanders at all levels have a simple and clear idea of how the Division would normally seek to conduct the defence. That may mean being able to act on information as slight as "The Division will defend Dunwhich by preventing the enemy crossing the Green River between Tucumcari and Anchester." This does not mean that this circumstance is likely. It means that in terms of training, should the Division commander give such direction, each unit commander can describe where they think they should be and what they should be doing. The defence, like the offence, is a totality of unit-level activities.

The Screen

The most often stated purpose of screening is to observe and report. It is most usually a cavalry or ground-manned reconnaissance task. It can be mounted or dismounted, but it must have superior or at least equivalent mobility to the enemy. It's a necessity that is generally terrain and enemy-type agnostic. You must be able to screen in the Arctic, jungle and desert. The screen is not limited to mechanised and armoured forces operating in central or eastern Europe.

The Screen is a catch-all term for the force that warns of the enemy approach and begins to degrade his readiness for battle by forcing him to conduct activities and expend resources, which would not be the case were the screen not present.

Beyond observing and reporting, the overall purpose of the screen is to conduct the covering force battle to prevent the enemy from being able to rapidly advance unopposed and not allow friendly forces time to dig in and prepare effective defensive positions. This means that the mission to screen the main force will, in all probability, migrate to being a highly mobile delaying action involving all arms. Thus, the term "covering force battle."

The lowest level of command this is likely to occur is brigade or formation, meaning it will be a multiunit task mission requiring combat and combat service support.

The mission verb "Delay" is often used to describe the covering force battle. There are various doctrinal definitions of "Delay," but they only make sense if the enemy is delayed enough time for sufficient defence preparation. This is not a statement of the obvious. Implied in many definitions of delay is the need for the delaying force not to become "decisively engaged", yet becoming decisively engaged may be necessary to buy enough time.

A screen generally comprises vehicles, dismounted observation posts, or both.

Ideally, a unit assigned to and trained for this task should be able to screen a 50km frontage in ideal conditions and to do so for 48-72 hours without resupply. This means that if everyone is static and well concealed, there is little chance of being detected due to administrative movement or resupply. Screening 50km demands the unit generate at least 40 dismounted observation posts or vehicles capable of the task. As of 2007, a UK Formation Reconnaissance Regiment equipped with CVR-T could screen 45km with 30 vehicles.[1] They need to be concealed against all forms of enemy reconnaissance.

This means 1,500m between vehicles. To ensure enemy reconnaissance does not approach or pass through the screen undetected, each OP should be able to detect vehicles and walking men at 750-800m in most conditions, which translates as 1,000m in terms of non-discretionary sensor performance. However, in most cases, topography, trees, buildings, and other obstacles will likely substantially decrease the on-paper performance of electro-optic or basic optical sensors. As previously stated, screening in mountains and jungles may be substantially affected by terrain that is impossible for either a man on foot or a vehicle to traverse, thus allowing the screen to concentrate on limited lines of advance.

If the unit conducting the screen is supporting a brigade of 3-4 Battlegroups it probably means the brigade has a frontage of no more than 30km, assuming a maximum of 12 sub-units (3 per battlegroup) and six are in the first echelon, with a frontage of 5km each. It might also seem logical to assume the screen is no more than 20-25km beyond the main defensive position. This probably requires a 40 x 40km or more training area for field training purposes.

The precise siting of each OP will be a major factor in the performance of the overall screen, and the only way to do this is for units and sub-units to get out onto real ground and conduct screens against live, non-co-operating enemies who will likewise try and detect them in return. Long-range optics, ground surveillance radars, unattended ground sensors and even acoustic sensors all have a role to play if sensibly employed. The ability of the OPs and vehicles to detect the enemy is a mostly technical function, as is the ability to effectively report sightings and outcomes of engagements.

Divisional UAS must also be trained as a means, along with multi-role radar, of detecting and targeting the enemy long before the ground callsigns are in contact. This will mean finding and targeting all enemy emitters, such as weapons locating radar, UAS control stations or teams and air defences to ensure overall freedom of action. It is fair to suggest that this could be considered a "deep battle" unconnected with the covering force battle, yet while the two are inextricably linked, they should be considered distinct and separate by battle space. Correctly done, this should avoid the need for coordination and deconfliction, thus cutting down transmissions and making the order process quicker and simpler. Deep fires and ISTAR should look beyond the covering force. The Close Support fires should be allocated to the covering force once the covering force is in contact. There will come a point where the battle space needs to be re-drawn but that should only occur after all the covering have broken clean and conducted a reward passage of lines. That timing needs to be decided by the Division.

To ensure the simplest and most robust conduct, the reward passage of line procedure should be as simple as possible and practised under high emission control levels and in bad weather.

Also important is the rehearsal and confirmation that the unit can move off the line of march or from a hide or assembly area and mount an effective screen at relatively short notice. Allowing for any distance to be covered to the screen line, a Divisional commander might reasonably expect some initial screening effort to be in place in two hours and for the screen to be bedded in and reporting at four hours.

The conduct of the covering force battle means being able to kill and destroy enemy reconnaissance, including armoured or mechanised fighting patrols of up to sub-unit size. This means that whatever comprises the screen needs to either be equipped with or supported by anti-tank guided missiles (ATGM) or a similar anti-armour capability. This has implications beyond the unit given the task and means that both combat support and combat service support elements, which may be attached to the covering force or screen, must be well-versed in the methods of operation required. This may mean ground dumping or caching ammunition and fuel for the screening force as it falls back. Caching complex weapons like ATGM may require additional thought and protection, as these are high-value ammunition natures that may be unwise to leave unguarded. If ground dumps need guarding, then the sustainment and administration of that guard force need to be accounted for in training and the staff work.

Rapid mine laying, bridge demolition, and the creation or improvement of covert swim and wading sites, using sub-surface bridging and bank preparation should the screen be withdrawing across a river or marsh area, may also be needed. If these things have not been practised and perfected in training, then there is little hope come "the day."

As should be obvious, the specifics of this are hugely context-dependent, but that does not excuse anyone from the necessity to be comfortable with this task, as with many others. The covering force battle is essentially a delaying action, but training for a delaying action requires the attention of the Divisional or even the Corps commander.

The Commander concerned must know how much effort the screening force battle requires regarding his time and attention. An ideal world would see that all Divisional commanders would have witnessed a great deal of covering force training from their time as either sub-unit commanders, various staff duties and almost certainly as a Brigade of Divisional Chiefs of Staff, so ideally, there should be a high degree of familiarity with tracking and resourcing the units and formation commanders involved in buying him time to correctly assemble the defence, if time were needed because it may be that the defence is in place and well prepared.

Under this circumstance, the primary aim of the screening force is to attrite the enemy's combat power down to a level where he cannot continue and, if sufficiently fixed, allow the counterattack force to be cued into action. The screening force can either seek to re-establish the screen or arrange a "relief in place" (RIP).

Far more likely is that the screening force falls back until it passes through the safe lanes in the main defensive position, where, hopefully, it retains enough combat power to regenerate to form a reserve and exploitation force. What degree of loss makes that possible will need to be considered, as will having the sustainment assets available to conduct the repairs and resupply.

As previously stated, this ideal will only be enabled by training where the units and sub-units concerned are confident in what they can and cannot do.

Some examples may be useful.

The siting and sustainment of the screen should be relatively straightforward to rehearse and test. Remaining hidden while observing is something that requires constant practice. Still, fortune is a foul mistress because the enemy might approach from an unexpected direction, or a flanking formation might withdraw as part of a wider plan, leaving an open flank. This means repositioning the screen or some part of it needs to be something everyone is comfortable with in terms of something that might need to be done at short notice and under less-than-ideal conditions. This might need to be done will maintaining a very low electronic signature or working within wider deception measures.

After 72 hours, the screening force may need to be withdrawn, so relief in place will need to be conducted, but that would require a considerable amount of vehicle movement and a large increase in signature, which might compromise the entire force, being able to walk forward replacement crews might be something to be considered. Maybe not the whole crew at one time to ensure a less dramatic change, but it would need to be trained for this to be deemed desirable. Not least because the recovering crews would be taking over the vehicles of those who replaced them, meaning commanding officers might find themselves with 50-80 percent of men they might never have met. That isn't a huge problem, assuming they know the relevant sub-unit and troop commanders. The argument presented here is not whether this could be done but what the implications would be if it were to be done. Almost all ground surveillance devices can spot moving vehicles at double to quadruple the range they can spot a walking man. That has clear implications for the conduct of operations. If the argument is that fuel, water and rations re-supply require vehicle movement, then the signature must be reduced as much as possible. How best to do this will require training and experimentation.

The actual circumstances and conduct of the screening force battle must also be subject to some rigour via field and other forms of training. There is a balance to be struck between destroying the enemy reconnaissance as they advance and imposing a delay on the enemy force overall. Therefore, training to conduct the screening force battle must allow for the greatest possible variations in enemy courses of action. The templated and stereotyped tactical doctrines of the Soviet Army, which give current shape and form to what the screening force battle looks like, may be very different from what occurs devoid of that context. For example, there may be no enemy reconnaissance. You might have to face MBT units moving forward in sub-unit columns. If the retort to this is, "Oh, but if they do that, they will have been destroyed by our deep fires/attack helicopters/wizard spells," then this begs the question of why any commander would not want to have considered that possible reality and what to do about it. The history of warfare overflows with statements of certainty as to things the enemy could never do yet did!

Again, this point speaks to the whole thesis of this work, which is not to imagine what future wars or conflicts may look like but to train to do the things which will gain you the greatest benefit regardless. Suppose your reconnaissance call sign commanders have trained for a circumstance where they may face an armoured sub-unit advancing rapidly and with little traditional caution, not because the enemy is reckless but because they wish to impose shock and surprise. In that case, those who have trained for that condition are better equipped for most eventualities versus those who did not.

The screening force must contain an effective anti-air element in terms of SHORAD to threaten enemy attack helicopters and UAS. This presents a dilemma regarding emitters, so until it becomes necessary to "go active", all anti-air detection should be electro-optical or even acoustic.

Large barrier anti-armour minefields are one obvious and useful measure to defeat rapid armoured thrusts. As previously discussed, buried AT bar mines tend to be substantially more effective because they can form part of a more carefully considered and integrated plan, which the covering force can use to its advantage. Being laid as part of an overall obstacle plan means that the minefields can be covered by observation, thus allowing any enemy attempt to breach the minefield to be targeted by fire.

So far, this discussion has assumed a mobile mechanised type of operation. Still, the screening force action must also be executable in other terrain where mobile, mechanised or armoured forces may not be the means the enemy employs. This would place a premium on dismounted observation posts, long-term ambushes, and command-detonated explosive barriers. Alternatively, or in addition, the enemy might be skilled at conducting formation-level dismounted infiltration along multiple routes. Regardless of specifics, this requires additional training and expertise from the units concerned and the command and staff responsible for the battle.

As previously stated in Chapter 8, anti-tank ditches are essentially anti-everything ditches. They should be three meters by three meters and probably enhanced with minefields and wire obstacles around and inside the ditch to prevent enemy infantry from merely scrambling across them.

The viability of constructing such obstacles in a contested environment must be considered, as must the terrain's topography. Training needs to account for the fact that the enemy may try to hinder both the laying of a minefield and the digging of anti-tank ditches.

Such levels of construction must be commensurate with the defensive advantage required with the primary aim being to force the enemy to concentrate to either breach the obstacle or to exploit the breach. Obstacle creation is costly and time-consuming. The more assets the enemy is forced to allocate to a breaching the greater their investment in that action will be. Mine ploughs and line charges are more common than the scissor bridging and bulldozers required to cross a properly constructed AT ditch. The location of anti-tank ditches and minefields should not indicate the location of the main defensive position (MDP). Ideally, they should be integrated within the covering force battle area to enable the covering force to operate with a significant advantage over a force without such obstacles.

However, this may not be possible given time and the tactical realities. It could take days to lay out an effective divisional obstacle plan, which should be done if time and conditions permit.

If time does not permit, tasking D9 Bulldozers to destroy the road network in the covering force area will suffice. This is done to prevent the enemy from being able to use the road network to support their attack. It also means the covering force needs to be sustained despite this condition. It takes about 30-60mins for a pair of D9s to render a six-lane highway impassable to any wheeled vehicle. The same capability can push large amounts of soil into small rivers and drainage ditches to cause local flooding, which impedes enemy progress. D9s can also collapse two to three-storey buildings into streets.

Understanding how to use such a capability to the best effect requires considerable study and experience.

Critical to this training is the use of obstacles to allow the covering force to break clean and recover through the MDP, where it can begin to regenerate both to form a reserve and to be ready to reestablish the screen once the enemy is defeated.

If a unit is screening on a 50km frontage, it is extremely unlikely that the whole screen will be in contact with the enemy. Hence, training needs to account for lateral withdrawal directions relative to the enemy's direction of attack instead of entirely retrograde ones. This has the added benefit of threatening enemy flanks and reserves and preventing enemy attempts to bypass the MDP.

The Block

The Block can also be described as the main defensive position (MDP) or the main line of defence. Regardless, the block aims to bring the enemy to a halt so that they can be defeated. The traditional stereotype of the MDP is that of infantry and armour units with two sub-units forward and one back, covering a 3-5km frontage with everyone dug in. The stereotype exists for a reason, and even the most obvious and templated defensive plan sufficiently well executed and trained can present a substantial obstacle.

Like anything inherently simple, the merit will lie with the degree of skill employed, and only from that basis can more advanced concepts progress.

Ideally, the enemy encounters the MDP as an ambush, though this is an ideal and should not be counted on. If the defence is to hold a key piece of terrain, then its location may be obvious to the enemy. The main criteria for the MDP, beyond anything else, is that it is hard to detect even if other circumstances have forced its location. All positions should be very hard to detect to ground and air observation and as electronically silent as possible. If it cannot be hidden, then a comprehensive decoy plan must be needed to ensure the enemy must expend resources and time targeting a wider defensive area than might otherwise be necessary. The action on the MDP aims to detect targets for indirect fire systems such as close support 155mm, mortars and grenade launchers. The aim is to deny the enemy a target on which he can focus his offensive support while subjecting his manoeuvre units to maximum attrition. This will best be achieved by observation posts forward of the MDP.

All this is easy to say but very hard to do. Any competent enemy will want to conduct a close target reconnaissance of the MPD but will hopefully be limited to long-range observation. Field training should allow both infantry and armour units to become skilled at concealment and the required routine in defence. However, exposing troops to long periods in field fortifications should be avoided. There are many problems, from boredom and lack of physical exercise to hygiene and welfare issues. Specialist observation teams or special forces have maintained themselves in long-term sub-surface observation posts for 14 to 28 days, but that was not in fighting a major defensive action.

Once withdrawn, those individuals often required no small amount of time to recover before being able to deploy again. Also, given limited training days in the year, how many does any Army want to burn to have its force sitting static conducting routine in defence beyond general familiarity?

This is a historic problem, so even in WW1, British and Commonwealth units only spent four to six days in the forward trenches and alternated between the same amount of time in the rear or combined with an additional similar period doing other tasks and training.

At the command level, this means that as soon as a defence is put in place, staff and command need to start working on how the units concerned will be rotated out of the front line. The process and procedures for that rotation are as important as those of the initial occupation. It is an identical problem to the screening force, so it can only be addressed by constant training and rehearsals. The training ideal is at night, in bad weather, on radio silence with little warning.

As previously stated, the actual conduct of the block should ideally resemble an ambush, but this is not without potential risk. It is extremely unlikely that it will be possible to engage the whole enemy force, so while the leading portion is subject to close-range, surprising, and shocking fires, most of the enemy force may be out of contact and able to react. Thus, the contemporary ideas for defence raise some dilemmas—the circumstances and particulars of how the enemy encounters or attacks the MDP are almost infinite. Still, it seems safe to assume that engaging the enemy across his depth, not just the leading elements, is required. This may mean manoeuvring to do so and withdrawing to avoid counterfire. Thus, an ideal is an enemy force decisively engaged by indirect fire, cued from undetected observation posts, allowing NLOS ATGMs and indirect-fire weapons to engage. What needs to be avoided is the enemy's ability to detect the position, withdraw and plan a prepared attack supported by artillery and additional forces manoeuvring to seek flanks and rears to find and defeat the reserve prepared for the counterattack. Thus, and as previously stated, the primary training aspect enabling this outcome is having a very hard-to-detect main defensive position. How is this done?

The simplest point to understand is that of the reverse slope. While simple, it is rarely applied and requires some additional context. The reverse slope is relative to the direction of the enemy threat. Its use is designed to prevent the enemy from being able to detect and thus target fighting positions from stand-off ranges. Fighting positions will most likely be detected once they are fired from or if poorly concealed in open ground.

Thus, fighting positions on flat ground overlooked by high ground must be positioned amongst or within buildings and woodland. The true meaning of the reverse slope is that no enemy force should be able to apply direct fire from their direction of advance at any range closer than your weapons cannot counter. In terms of almost any handheld weapon, that is probably less than 100m. While the reverse slope is the literal use of the reverse slope, it has a wider and more universal application in terms of not allowing your position to be dominated by stand-off distances. Looking back, the reverse slope you are sited on will be the forward slope from some other perspective, and this should be the location for ATGM and other direct fire systems which can engage the enemy on the crest, within your position and on the flanks if they attempt to bypass.

The next item to consider is dispersion. While the fundamentals of defence against a mechanised, armoured or even dismounted enemy have remained pretty much unchanged since about 1918, the distances at which those fundamentals are applied have increased beyond recognition for all communications, sensors and weapon systems. That said, the performance of the human eye and the body's capacity to move at speed carrying weight remains mostly unaltered, bar the benefits of modern health and nutrition. While the MDP is designed to halt the enemy, it is more specifically intended to halt him on a particular line of advance or route. Ideally, it should stand between him and his objective, but the MDP is not a coherent and cohesive linear array of fighting positions. A platoon might have a 400m frontage and 300m depth. A Soviet Motor Rifle Platoon did, but this was a doctrinal template contrived for training on the Russian steppe and not subject to the vagaries of terrain and intervisibility but modern weapons and sensors performance more than allow for a well-equipped and well-trained western platoon to cover more than that area if terrain allows. While a platoon could cover 1,500m using enfilading fire, dispersing a platoon over such a distance would require considerable thought and practice to confirm issues such as casualty evacuation, administration, and resupply.

Even communications could become a problem, especially if limited to landlines in a zero-emissions environment. It would be perfectly possible to allow for 1,500m between platoons, assuming that distance was adequately resourced with the types of weapons and sensors stated. Allowing for two platoons forward, a sub-unit could cover greater than 4,000m. A unit could cover more than 7,000m. These are levels of dispersion far over most thinking to date and are unlikely to be effective or understood without sufficient training. Were that the case, then conceiving of a unit holding being tasked with 6-7km frontage and 5km depth is within the bounds of reason. Detecting or targeting any part of that array would require the surveillance of 30-40 square kilometres with platoons and sub-units dispersed far beyond what current artillery systems can effectively deal with unless precision fires target individual vehicles and fighting positions.

The secondary effect of such dispersion is that it makes it far more likely that a great deal more of the enemy may be within the position than was previously likely, adding the idealised ambush effect. This leans towards the best outcome: the enemy is destroyed and defeated within or even behind the MDP, making it far less likely that any element can withdraw to attempt a second attack. Likewise, dispersion allows for laying protective and barrier minefields immediately in front of the MDP and even within the MDP itself without necessarily nullifying the possibility of such minefields negating counterattacks or armoured and mechanised manoeuvre. The MDP should never be established in open ground, which allows armoured vehicles to drive unimpeded over fighting positions, but in terrain that the armoured fighting vehicles can access, such as villages or areas of bush and brush with widely spaced trees, anti-armour mines within the MDP should have a decisive effect. Again, this would require a high degree of training.

Ideally, the enemy should never reach the reverse slope to be targeted by direct fire, but this can never be guaranteed thus the necessity for the reverse slope position.

The Counterattack

On a purely technical level, a counterattack restores the defence, or a part of the defence threatened or lost. The reality is several actions and activities might collectively be considered under the umbrella term of counterattack. If the block is designed to stop the enemy, a counterattack will defeat him and exploit the high levels of attrition and loss a successful block may have delivered. Doctrinally, most armies have suggested most levels of command above that of platoon or troop maintain a local reserve, but that can vary substantially in size and makeup. That may be a sub-unit or just a platoon at the unit level. Regardless, the most basic purpose should be that if the enemy attack is making headway, the reserve executes a counterattack against them and restores that part of the defence. The more complex idea is where the defence is predicated on executing a counterattack or counterstroke; it is planned and organised with a force detailed for that purpose and holding other levels of reserves. It might be that the covering force identified the main thrust of the enemy advance and can cue an armoured or mechanised counterstroke into the enemy flank or rear. This could equally be an attack aimed at the enemy force while preparing for an attack, commonly described as a spoiling attack. All these actions can be conducted by manoeuvre, direct or indirect fire and a combination of all three. Indirect fire, as in artillery and long-range ATGW, must be controlled by the force conducting the counterattack for obvious reasons. This means that control needs to be transferred at some point and by someone authorised to do so. Opinions on this may differ, but as with all written so far, this action needs to be trained for, and how it is conducted needs to be clearly understood.

The qualifying characteristic of the counterattack or counterstroke is that the defence is predicated on actively manoeuvring against the enemy and not remaining static, locked onto a piece of terrain. Counterattacks against a mounted armoured or mechanised enemy will most likely require armoured or mechanised counterattacks, raising the issue of where to hold and conceal the counterattacking force or mounted reserve. While counterattacks are most obviously associated with manoeuvre, there is no reason why the same effect cannot be delivered by fire. The objective has been reached if the net result is an enemy force that is so damaged that it cannot move, recover, or retreat.

Regarding training, the key objective will be for the relevant commanders to know when, where and under what conditions the counterattack is launched and, equally important, how to generate and organise forces for that purpose. It is too simple to state the necessity for a reserve without clearly defining what that means in practice. As previously stated, if the counterattack is the key to the defence, then the forces allocated to it are not available for the unforeseen circumstances that may require the employment of the reserve. There can be two views of reserves, both common to defensive and offensive operations and not necessarily in disagreement, but the first would be an uncommitted force available to address a point of threat or crisis, and the second would be a force that you intend to commit as the decisive act of the operation.

It would seem perfectly possible for the nature of the reserve to alter with the level of command it is held at. It may be that at the unit level, a platoon suffices to fulfil the requirements for a reserve. At the formation level, it may just as likely be a sub-unit as a unit, and at the level of a division comprising two brigades, one brigade may conduct the screening battle and block while the other brigade is the counterstroke force. In terms of training, there must be a high degree of confidence that the counterstroke force can reach its jumping-off point or line of departure undetected and unchecked. The counterstroke force will almost certainly have to have been in several separate hides to reduce its signature, to when and where it assembles will require a solid and simple plan. It must also be recognised that the counterstroke or attack conditions may be unclear or ambiguous at best. Command and staff training must account for conditions where the enemy is engaged inside or in front of the MDP, but the situation is unclear. While a counter-attack might be launched to restore the MDP, the optimal condition for the counterstroke would be an advance to contact into the attacking forces' deep flank to defeat the attacker's reserves and threaten or attack his combat support echelons as in finding and killing his close support artillery. The conditions that separate these two courses of action may not be obvious when you want them to occur, so training needs to differentiate between when one is applicable and when one is not.

A counterattack may be a relatively small force launching on an unambiguous set of conditions as a "be prepared to" (BPT) task. It may have two or three BPTs, all of which have been rehearsed. A counterstroke usually requires a far larger force aiming at more decisive results. Almost every defensive plan will compromise both counterattacks and counterstrokes, but resourcing both equally will diminish the overall effort. One way to address this is to use the covering force to track the enemy advance to cue a counterstroke into the enemy flank. This may be the only course of action open until the MDP is prepared, in which case orders and training must prepare the force for conditions where the units comprising the counter-stroke force are held ready behind the screen as it is established. In terms of priorities, this makes good sense. This is only possible if the entire Division already has clearly defined roles, as alluded to in the opening of this chapter.

This is not a particularly complicated or sophisticated idea. Still, it can only be exercised at the Divisional level of command, and the frictions associated with making it work require considerable real estate or training areas.

The Counter stroke should be conducted as an advance to contact (covered in the next chapter) with a clearly defined limit of exploitation.

Recovery and Sustainment

It is important to realise that a defensive battle against any competent enemy will not be one event of limited duration, after which the enemy is defeated, and the war is won. The enemy may continue attacks for months or even years. Due to exhaustion, attrition and breakage, any force must be rotated out of its defensive position within 12-14 days or maybe less. It must then march to its hide area and begin regenerating. Training must account for the conditions under which that will happen, even if it is just command and staff study periods. Such matters include units reporting an exact state of manpower and equipment readiness to the division or higher—a time estimate for when they will be returned to a partial or full operating capability. Once this has been achieved, there needs to be a plan for the unit to get full periods of relaxation. Military relaxation must be planned and controlled to avoid mass disciplinary events such as riots and fighting between units. Anyone surprised or alarmed by this has probably never spent much time around soldiers. This may seem more than obvious to others, but unless carefully considered and developed into solid guidance, all forms of misfortune might occur. Soldiers need repair in the same way as equipment. Like equipment that is best done by collecting them into one place to make the best use of resources so that they can be physically and psychologically regenerated to return to operations as soon as is practical or before in the case of extremes.

Conclusion

To state that the defence requires training is no revelation, but that training must be conducted with the backdrop that the defence will provide a stepping stone to future offensive and, thus, decisive operations. The mythology of the Cold War's "Die in Place" cannot take hold, so every defence aspect must be thoroughly rehearsed for, planned and discussed to produce the most beneficial training.

Endnote

1 Junior Officers Tactics Course Handbook, 6th Edition 2007.

13

THE OFFENCE

The purpose of the Offence is to change a situation or condition. You seek to destroy (thus defeat) the enemy or capture some place or terrain feature. Offensive operations seem to be trained for far more than defensive ones, which may not seem surprising given the need for activity, imagination, and overall engagement. Politics, rather than warfare, tends to demand a need for offensive operations as being those most likely to gain the outcomes desired. Regardless, training for offensive operations tends to provide the bedrock of ideas and methods which have far-reaching application to the force overall.

In general, the modern offensive should seek to advance on a broad front to fix the enemy across his entire frontage while simultaneously detecting or creating gaps to feed exploitation forces able to conduct attacks on support echelons and reserves. This is no different from 1918 in terms of a mechanism to inflict defeat. The key difference is that thanks to modern communications and sensors, even a small force exploiting enemy depth can leverage large amounts of firepower. The range and speed of almost all vehicles far exceeds those of even WW2.

Reconnaissance

The necessity and value of reconnaissance cannot be overstated. While seemingly obvious, the separation of reconnaissance into some specialist skill set, which makes it the preserve of only a small part of the force, has almost no merit. Infantry units have tended to place their brightest and best in reconnaissance platoons because of the demand to read maps, operate radios, and produce accurate verbal and written reports, which may have been somewhat beyond most infantry battalions' training schemes. The reality of many modern infantry operations is that dismounted reconnaissance is inextricably linked with modern offensive operations.

Reconnaissance is likely to be what many manoeuvre units spend their time doing when it comes to offensive operations for the simple reason that, as we have previously described, for defensive operations, the ability to hide and remain concealed is one of the key enablers of modern defence. To attack, you need to find a modern defence.

It is an overstatement to say, "Everyone does reconnaissance all the time", but the statement attracts useful insights even when rebutted. The planning for all offensive operations is predicated on where the enemy is expected to be relative to the objective.

Equally important and maybe more so, is where the enemy is not. While commanders will have access to many intelligence and information sources as to where the enemy is and what he is doing, very few of those have implications for the training of the force under his command. If you are a brigade commander, the fact that your staff has access to Corps-level ELINT- while extremely important - is unlikely to impact the formation training program as in the need to exercise the brigade in locating an enemy both on and around the objective.

Thus, all offensive plans require a terrain objective. That may be a specific locality, feature or just a line on a map, as in a control measure. The mission to "Capture spot Height 178 by 05:00hrs" is conceptually simple and represents the baseline requirement.

"Find and destroy all enemies within Ops Box Manta by 18:00hrs on the 23rd" is far more complicated given that Ops Box Manta is 15 x 15km and the 23rd is two days away[1].

The discussion that follows must necessarily span ideas useful to both. In all cases, it is important to realise that the purpose of reconnaissance is directed towards gaining the mission objective. Land warfare requires physical surface-based methods of reference on which to base offensive concepts. Reconnaissance uses this as the basis for such a mission. This is generally distinct from the protective activity of screening and surveillance, which protects the force while it is static or preparing its defence. This has considerable relevance because whatever level of command is concerned, it must be able to protect the force. At the same time, other elements conduct the reconnaissance missions on which its next operation is based. The requirement for any level of command to conduct reconnaissance generally comes from the level of command above and the requirements of the campaign plan or wider operation. A unit, formation, or division all have the same need to find the enemy in areas X or Y. This may be of a massively varying nature and context.

It does not help specify or describe the training needed to prepare the relevant command level, but it is nonetheless insightful and important. Finding the enemy is hugely important. So much for the transparent battlefield.

Offensive operations work backwards from the need to detect the enemy in the same way defensive operations do, so the immediate training requirement is for staff and commanders to generate orders that give their subordinates simple and achievable tasks and missions. Those tasks and missions must axiomatically be aligned with the levels of training already achieved. Can a platoon detect and report on what, if any, enemies are at the farm at grid 123456? Can they do it without the enemy being aware? Can they leave a section to maintain that location under observation until H-Hour? If they see vehicles on the objective, can they identify the type?

Almost everything about training a platoon, sub-unit, or unit to conduct reconnaissance will lift that organisation's overall level of training to a level not achieved by any other type of mission. At the unit level, it is thus easy and useful to combine sub-units conducting routine in defence against sub-units or platoons conducting reconnaissance. A wider capability can only be created from a confirmed dismounted low-level skills base. The same is true of mounted, as in the use of vehicles and Unmanned Air and Ground Systems (UAS/UGV), and here, a note of caution needs to be applied. Recent conflicts have seen the wide employment of cheap commercial UAS for reconnaissance by forces with varying levels of training and limited force-wide capability. Still, their existence does not reduce the requirement for a manned mounted and dismounted ground reconnaissance in terms of training.

Allied to infantry units being able to conduct reconnaissance is the formation reconnaissance units mostly predicated on the cavalry capability as in a mounted reconnaissance capability with the option to dismount forces for specific tasks. Mounted reconnaissance moves from one observation position to the next as a pair or multiple callsigns along a predetermined route to reach a known point. The aim is to detect the enemy before it is detected. However, the more likely occurrence is being detected and fired upon, but hopefully under conditions where good use of ground and cover from other vehicles reduces the likelihood of casualties. This possibility leads to the conclusion that manned reconnaissance should be conducted by vehicles with very high levels of protection. Larger, heavier vehicles can have less choice regarding the ground they can use for their approach. This very demanding form of training is the most likely to build the skill set required for real operations, hopefully providing a context where theatre and Corps assets will have provided a better picture of the likely enemy lay down against which to operate. While this is ideal, it would be foolish not to work within scenarios more demanding than those likely to be encountered, where no enemy location is known.

As previously stated, the modern offence needs to detect the modern defence, which is designed to be hard to find and aims to detect the adversary's reconnaissance force.

The Offensive Deep Battle

As previously discussed, the aim of the offensive deep battle is the preparatory operations to destroy the enemy air defence to allow freedom of action of air assets to locate and destroy enemy C2, as well as the more traditional targets such as logistics. The Deep Battle is best conceived of in terms of units and resources allocated to battle space in the simplest possible form to reduce the need for coordination. The deep battle has a clear demand for deep reconnaissance, which will almost certainly reside with Corps or Theatre assets held above the Division. This means the Corps or Theatre may well fight their own deep battle that enables the divisional deep without the Division having to expend resources. The Air Campaigns in both Gulf Wars demonstrate this, but this doesn't give many insights into training the Division's deep battle assets. The Division's deep battle is primarily counter-battery, but a high-value target list should be maintained by the Division's deep targeting cell. Two criteria qualify high-value targets. These are time and place. What becomes a high-value target will alter over time and proximity because striking deep targets usually consumes limited resources. No one should just be striking every high-value item they detect unless it is part of a plan that accounts for the expenditure of resources to do so. One of the most fundamental tasks for depth fire assets is destroying or suppressing enemy artillery, air defence and UAS control stations and personnel, so this should be a primary training objective of the Division's deep battle complex for most operational conditions, assuming that higher command is creating the overall freedom of air action. After all, your air force's primary task is to destroy the enemy's air defence systems to enable the overall counter-air plan, enabling deep interdiction. Further discussion of the Air Force mission lies outside this work. Suppressing or destroying the enemy artillery, air defence or deep attack capability does not necessarily mean targeting fire platforms such as rocket launchers and guns but killing their targeting sensors, C2, EW and networks. This is not to say killing platforms does not matter, but their effect is limited compared to striking a command staff and air or weapons locating radar.

The Simple Attack (Knowledge)

The defining characteristic of a simple attack is knowing where the enemy is and what they consist of so that you can plan and execute with a high degree of confidence, assuming you have the resources you need. Successful reconnaissance generates the conditions and information, describing something as a simple attack.

Assuming successful reconnaissance for training purposes, the simple attack is primarily procedural in terms of things that can be achieved in training to create wider benefits on actual operations. By far, the most important element of this is the artillery fire plan and how that will be coordinated with the various assaults in terms of timings and sequence.

For discussion, let us assume a light force infantry battalion attacking a dug-in infantry sub-unit. The fire plan will work backwards from the known enemy locations and what can be confirmed by direct observation. For training purposes, this is not hard to rehearse and practice. However, you are still faced with the contradiction that a simple attack, as defined here, is only possible because you know where the enemy is, which is likely to be rare in the real world. That said, every infantry battle in the Falklands War of 1982 was conducted against known and, in some cases, well-mapped enemy positions.

In every case, this was due to extremely skilled and diligent reconnaissance activity.

The training objectives for a simple attack are simple and measurable. From receipt of the order from the brigade to the first sub-units, crossing the line of departure should ideally take no more than four hours, with the battalion completing the writing and issuing of orders in no more than one hour and twenty minutes. The fire plan should be within the capacity of the gun regiment in support, as it only demands things they have proven capable of in training and the relevant gunnery practices. Forward observers and Fire controllers are allocated to their various sub-units and can maintain effective communications with the gun and mortar line to adjust fires as required. Traditional success then requires that all sub-units move into their assembly areas, onto their lines of departure and move forward per the fire plan. Modern success means moving from your hide location to a platoon control point from which you will depart on a bearing towards the enemy because enemy sensors will detect and punish any large concentrations of troops.

This is as much as can be trained for via process and procedure. It all needs to be accomplished with the minimum of signature to ensure that when the attack occurs, it is both shocking and surprising to the enemy as it concerns time and direction. Surprise is something the enemy is unprepared for, and shock is an inability to make effective decisions or conduct actions because you have been subject to physical or psychological harm. If a unit can consistently do these things quickly and violently, they will have accrued a valuable and useful skill set. A high skill level means success still occurs when friction is injected into the system. Friction can and should be everything from less preparation time to enemy counter-battery activity disrupting the fire plan or wide area enemy communications jamming. The other major source of friction would be the requirement to end the attack in the defence. Unless planned and conducted as raids, all attacks end in defence, and this is particularly necessary if the enemy has a strong reserve, which needs to be accounted for. Thus, completing a successful simple attack establishes a defence beyond the objective designated for the mission, prepared to meet any counterattack.

The Assault

Special attention must be paid to the assault or that part of offensive action where sub-units individually or collectively fight through and defeat the enemy on their main defensive position. In its dismounted form, the assault is characterised as the subunit moving onto the enemy position after sufficient suppression and destruction have been applied by direct and indirect fires to allow the attacking force to move to a point where it has such proximity and numbers that the enemy surrenders. If the enemy continues to fight, they are defeated with small arms fire and close-range projected high explosives as the assaulting infantry clear through the position. Training for this needs to be relentless and thorough.

The mechanised or armoured assault requires the attacking sub-unit in APCs or IFVs to move from the dead ground or cover where it formed up across the open ground towards the position and aim to debus their infantry either on the forward edge of the position or seek to drive into the position. There are some variations on this theme. The whole time the sub-unit moves, the enemy is hopefully suppressed by artillery or direct fire, as are any depth or enfilading ATGM positions. A lot of training is focussed on this event, including live firing. It could be suggested as being intended as the decisive act for which all else must be subordinated.

Regarding training for war and the conduct of the assault, several issues here are worthy of discussion. The first is that unless the enemy has been sufficiently demoralised by the preparatory fire, the assault is unlikely to succeed. The second is that a dismounted force is far more likely to gauge how successful the preparatory fires have been than a mounted one. Additional fire missions can be requested if they receive effective fire while moving forward. An armoured or mechanised force may lack that awareness until being struck by anti-armour weapons. Once across the line of departure, it is most likely committed to moving onto the enemy objective. Ultimately there are an almost infinite number of conditions which bear on the likely success or failure of a mounted assault.

Still, the overarching factor in success is either the effectiveness of supporting preparatory fires or such a degree of shock and surprise in terms of the assault's speed and direction that the enemy is overrun before they can react. Given the noise and signature of armoured and mechanised forces, this cannot be relied upon but may not be impossible to exploit should circumstances allow. This means being skilled at very rapid assaults against enemies probably only detected at close range and probably best executed as a contact drill rather than a planned activity using some form of battle procedure.

As explained in the previous chapter, it is trivially simple for even poorly trained troops to dig deep and surround their positions with high-density anti-tank minefields, which can be installed in less than 24 hours. Even a low-density or nuisance minefield can close off a likely approach for enemy armour.

Thus, ideally, the assault, which is so much trained for and focussed upon, should never take place. This may be aspirational in the extreme, and it is all too easy for discussions to talk about "bypassing" to "get into enemy depth", but then the by-passed positions react to block re-supply, reserves and evacuation. The assault may be non-discretionary but must be executed using effective preparatory fires combined with shock and surprise from deep flanks or rear to the enemy's expected threat direction.

The Advance to Contact (Ignorance)

The advance to contact could be defined as one where little to nothing is known about the enemy. It would be an advance towards an objective, with little or no confirmation of where the enemy is or in what numbers. Thus, an advance to contact. A broad doctrinal description of the advance to contact is an operation that seeks to re-establish contact with the enemy or to contact an undetected enemy, so nothing about this idea is a radical departure from widely accepted practice.

In training, a platoon or sub-unit moves through an area and contacts the enemy under varying conditions and circumstances. Still, in all cases, they must be attacked and defeated. The advance to contact is a well-established training vehicle, especially for small unit commanders in infantry battalions. Thus, the advance to contact is ideal for platoon and company training because it requires rapid decision-making, clear and simple orders, and high levels of initiative by junior leaders and expertly controlled close supporting fires. The circumstances are deliberately chaotic because unit commanders want to know who can work within warfare's chaos and who needs more training because they cannot.

A unit, formation or brigade trained to conduct a wide frontage advance to contact against a peer competitor is likely to develop a high level of skill in offensive operations, bearing in mind the conditions previously described in training for the defence. The rapid decision-making, clear and simple orders, and high levels of initiative required for sub-unit and platoon are now elevated to all unit commanders and the staff. This logically places a high demand on drills and the rapid execution of immediate actions. Still, it also places high demands on sustainment and what is required to re-organise and regenerate in terms of sustaining the required rate of advance with the threat of being subject to counterattack, where the formation is suddenly faced with an encounter battle, which may mean switching to the defence. As with the simple attack, all offensive action should allow a rapid transition to defensive action to deal with unforeseen conditions. In the case of the encounter battle, this will probably only be a proportion of the force.

Modern conditions must recognise that any advance by an armoured or mechanised unit will probably be detected by enemy sensors, thus allowing it to be targeted by various means. This means that the advance to contact should consider dismounted infiltration as a primary means of getting sensors and communications forward while remaining below the enemy's detection threshold. A platoon can advance 10km forward during one period of darkness, and likewise, it is just as easy for nine platoons to advance on nine routes across a 12km brigade frontage. This is easy to train for. Each platoon conducts a fighting patrol while supported by formation or divisional assets. Platoons are intended to detect or bypass enemy contact. There will be various methods of sustaining those platoons and ensuring the same action can be repeated across multiple periods of darkness to allow the formation to advance with as low a signature as possible. The wider application of this methodology is context-dependent. Still, it is easy to train for and can be done on public and private land with little likelihood of damage to infrastructure or crops. Success or failure for this training is identical to reconnaissance, as in moving across an area and detecting the enemy before being detected. An even greater success would be to pick a route of such obscurity and difficulty that the platoon is never detected and completely bypasses all enemy locations, enabling it to reach the enemy rear areas and call for fires on high-value assets such as C2 locations and unit echelons. Again, it needs to be emphasised that this is not necessarily a concept of operation. Still, it is a training scenario which builds a high degree of skill and capability because it is inherently challenging. It also provides the basis for a concept of operation should that be seen as something likely to be successful given appropriate terrain and an enemy against whom this might work. Wide area unit-level dismounted infiltration should be an arrow in the quiver of any competent infantry force.

The Encounter Battle

Encounter battles are historically rare in that a collision exists between two moving forces engaged in offensive action. That does not excuse or suggest not training for such an event. An encounter action consists of an initial block immediately followed by a flanking attack as might be conducted against any enemy location during a hasty attack during an advance to contact. The main difference is the speed of reaction required, thus the lack of supporting preparatory fires and the flanking force having no clear objective compared to an attack against a known location. Training for such an event is expensive and for anything above sub-unit but can be reduced to its parts to gain useful proficiency. The instrument of decision for the encounter battle is most likely to be the flanking action so either the immediate action or rapid planning and orders that create that are most likely to be the item that gains success. This can be treated as a discrete unit or sub-unit level action for training purposes. Trained for in open ground, there is almost no need to conduct this as training bar a discussion as to what the actual scheme of manoeuvre should be like as concerns time and distance. Move the training to real terrain where there are woods, small streams, rivers and even railway or road cuttings and embankments, and the rapid flanking action slows considerably, further retarded by having to guard against another chance encounter with an enemy, perhaps attempting the same. As with any training, this means a unit, sub-unit commanders and platoon leaders becoming well acquainted with the vagaries of real terrain, which, while common to all tactical training, will be exacerbated under the conditions of the encounter battle where speed of movement and reaction is paramount. Dismounted forces can access and move through terrain impassable to mechanised or armoured forces. They cannot do so at the speed needed to out-manoeuvre mechanised or armoured forces. Dismounted forces may remain undetected by the enemy forces depending on terrain and might stop or degrade mechanised and armoured forces with hand-held or guided anti-amour weapons. Sufficiently well-trained, dismounted forces can attack the mounted enemy in close terrain, offering the potential to halt or blunt any attempt at a decisive manoeuvre. This would, in part, strongly suggest the need for infantry and even cavalry forces to seamlessly transition from mounted to dismounted manoeuvre. There is obvious merit in a force not being blocked by a rock-strewn river, which, while impassable to any vehicle, might reasonably be traversed by the men in those vehicles. The requirement for this is (or not) most likely to be demonstrated by training that accounts for being able to move tactically across broken terrain.

River Crossing

River crossings should be a discrete operation focussed at the formation level, requiring several separate units or detachments to succeed. It is normally considered an enabling action not directly associated with an offensive action but is done here to examine some wider elements of the offence. Any river crossing must allow for a comprehensive fire plan in terms of what is currently described as general support and deep fires. In terms of training, this does not affect the actual conduct of the crossing itself. Still, the sustaining and movement of guns and fire support platforms requires a necessary amount of staff attention and rehearsal.

There is no such thing as a standard river or a standard approach, exit or bank, so the simplest and most robust process, procedures and equipment choices must have the broadest possible application. Regarding training, there are considerable safety constraints, at least equal to those of live firing. The training for river crossings generally comprises three distinct forms of activity:

- The securing of the bridgehead
- The construction of the crossing
- The crossing procedure.

Ideally, all three are conducted as part of field training as the formation crosses a river as part of an advance, but the training reality is that short of that, all the moving parts will need to be perfected separately.

Securing the bridgehead will normally be a dismounted infantry task, meaning unit or sub-unit training in safely approaching and crossing a water obstacle as part of a dismounted advance. This may well include the conduct or inclusion of the specialist reconnaissance capability associated with combat engineer crossing of rivers. That said, the requirement for specialist engineer reconnaissance can be overstated. The skill set required to assess a crossing site can be proliferated to other arms and services via training and is not the sole domain of the unit that builds the crossing. Infantry units will require support from boats as swimming, even well-trained troops, is rarely, if ever, practical. A unit equipped with amphibious vehicles will thus have a considerable advantage over one without. A dismounted unit usually needs to secure at least a 5km x 5km bridgehead on the far bank. This may involve a substantial amount of fighting and counter-battery fire, especially as there is a clear need to nullify the enemy's ability to interdict the crossing.

Given the time constraint of one period of darkness, this may be a considerable challenge and needs co-ordinating with the construction of the crossing site in case concurrent activity is required. The reality of almost all crossing operations is that if the bridgehead is secured during darkness, most crossings will occur in daylight unless multiple crossing sites ensure a very high rate of vehicles moving across the river. The number of crossing sites, dispersion, and nature will be discussed elsewhere. Still, in terms of equipment, the most likely solutions are pontoon bridges, rafts, or varying combinations of both given ribbon-bridging and specialist pontoon vehicles. Training the relevant unit to construct, operate and maintain a crossing site is something inherently simple in terms of measurable conduct and success. It should not be overcomplicated.

The complicated part that requires the most training and rehearsal is the crossing procedure, but the crossing procedure can be done without an actual crossing. It is essentially a traffic management problem needed to mitigate the threats concerned with indirect fire and signature management. The required skill set applies across a range of operations beyond river crossing. Each unit needs to be broken down into vehicle packets, which move down a series of hide sites, waiting areas and checkpoints, culminating in the crossing of the river and being released onto the far bank, where they will move directly to their deployment position or hide site, in compliance with routes and traffic plans which maintain dispersion and manage signature. The entire process needs one point of command to manage the choreography required. This must also be done with a minimum of voice or data transmission. Training for this can be done almost anywhere in the actual conduct of the crossing or crossing sites, in the car park where the vehicles concerned must wait for a specified time. If the vehicle packets arrive at the right time, in the right order, and depart the site onto a clear route, then the reality of the bridge is just something to be imagined—the bridge or rafts may be no more than traffic cones and tape. This is simple and easy to do.

There is no reason why an entire divisional river crossing cannot be rehearsed and drilled with two to three infantry battalions conducting a light force crossing on some stretch of water, ideally a river, an engineer unit conducting a bridging exercise, using the same area, and then a whole divisional movement using multiple training areas to simulate multiple crossing sites.

At least one of the multiple crossing sites will have to allow the movement of casualties and vehicles back across the river. This is far less of a rafting operation problem since rafts need to make return journeys, but it will need accounting for using single-lane crossings, so it still needs planning and practice.

Ideally, river crossings become trafficable routes where the actual crossing is a single-lane bridge, which does not impact the convoy speed of the vehicle packets. Still, this should not be counted on, given the advent of long-range precision weapons able to target such equipment.

Obstacle Breaching

Obstacle breaching is mostly associated with offensive operations. It is not impossible that in extremis, breaching might occur under other conditions, but the training should suffice for both. Obstacle breaching is not limited to the Middle East's mechanised and armoured battlefields or, the War in Ukraine. Such obstacles can easily occur in a forest, jungle, Pacific, Mediterranean, or Caribbean Islands. They can comprise multiple rolls of razor wire interspersed with anti-personal mine or unattended ground sensors cued to pre-planned artillery fires. Very few terrains and conditions absolve your force from being proficient in obstacle breaching.

There is a belief that the apparent similarity between a river, an anti-tank ditch and a minefield in terms of linear arrangement somehow makes for a templated solution that sees all three as essentially the same problem. This is simply not true. It is not inconceivable that all three could form part of one obstacle line, but there are more obvious problems. The existence of the river and your preparation to cross it is obvious.

There are no unmapped rivers on the planet. A minefield may come as a complete surprise, as might an anti-tank ditch, though that is unlikely. The time and effort committed to finding minefields are very different to that of rivers, thus impacting staff planning and how the staff see the problem. Again, the artillery fire plan will be a critical and central part of the operation because of the need to suppress the enemy artillery, which should be covering the obstacle, assuming they are well-trained. However, the weight of training will fall on the technical means of creating a safe lane across the minefield, which may well be >1,000m deep and crossing any anti-tank ditches. Those technical means will not be discussed in detail here, but the tactics, processes and procedures common to all breaching operations generally fall in the following areas:

- Action on the obstacle
- Deployment to breach
- Breaching
- Exploitation.

Action on the obstacle should mean that once an obstacle has been discovered, training accounts for several things, the most important of which is the fire plan needed to either cover a withdrawal if required or the further actions needed to breach. Obstacle breaching can be deliberate or hasty so that the broader idea can conform to the simplest possible methods of obstacle breaching may be critical. A simple attack would assume the obstacle has been detected and that the breaching action is part of the orders and something everyone expects. This may well mean encountering a minefield or obstacle which was either not detected until contact or rapidly laid by remote means such as artillery and rockets or scattered from vehicles and helicopters. Again, the need to move to the maximum level of dispersion, which is practical, might ensure that only a small proportion of the formation or unit is affected by any indirect fire threat be that conventional or remotely delivered mines. Historically speaking, the most likely current threat is buried anti-tank mines. While barrier minefields can be several kilometres in frontage and, as previously stated, several hundred meters or more, deep, smaller nuisance minefields are an obvious option for most enemies seeking to constrain freedom of action and impose delays, so the simplest breaching action may be to just drive around the minefield having ascertained its extent. This is not as obvious as one may suppose and needs to be trained for in that all sub-units in the first echelon, the reconnaissance force, and flank protection all need to understand that getting forward and around the obstacle is something everyone needs to do. Staff need to be trained and prepared to alter unit and sub-unit boundaries to account for this. Concurrently, the breaching force should move forward to act on the obstacle if required but should also be kept under cover until the last safe moment. This is not merely an obvious doctrinal statement but something that has to be trained for and discussed by the relevant commanders. Training needs to account for the judgement that where you contact the minefield or discover the obstacle may not be the same place you decide to breach it. While seeking to circumvent the barrier and deploying the breaching force may well be concurrent, there needs to be one level of command that can coordinate these as two distinct actions. This is because the resources needed to conduct a breach should not be expended if not necessary. It is possible and historically valid to encompass a condition where some vehicles find routes across a minefield due to poor density or planning. They can easily traverse ditches or remove obstacles for the same reasons.

The breaching action, as in using an explosive hose, thermobaric carpet, mine ploughs and rollers, are all discrete technical actions best practised in specialist range areas. It may seem blasphemy to suggest that the breach can be relegated to a notional action in most training events. Most armies will not allow the use of explosives without considerable safety constraints. Skills will likely become dismissible routines if you only practice unit and formation-level breaching on highly familiar terrain.

Assuming a successful breach, the methods by which that breach is exploited needs to be well understood. But before discussing that, training needs to address the remedial action in case of a failed or only partially successful breach, as in hasty breaching action that needs to be halted due to casualties. A deliberate breaching operation needs to be planned and executed. The unit-level battle procedure must account for this in transitioning from one to the other.

Exploitation is about passing the platoons and troops in the first echelon through the breach as fast as possible and using that freedom of action to create additional breaches or crossings if required. In context, if a ditch is covered by a 23m scissor bridge laid by an AVLB, the rate at which vehicles cross that may be as slow as 1 per minute in the dark or thick hanging dust from artillery, and the spacing required for sound tactical dispersion. It will seldom exceed three. This means that a 130-vehicle battlegroup will usually take between 70 minutes and two hours to cross an obstacle with one bridge.

Ideally, you need at least three breaches or crossing points, and ensuring the even distribution of traffic to each requires constant rehearsal and practice. Suppose a formation advance is contingent on a reconnaissance unit being fed across a breach. In that case, the intricacies of the timings will impact the rates of advance and the likely enemy reactions. None of this is particularly complicated if the relevant staff and commands have practised it and are familiar with the most likely first and second-order effects of the delays and problems breaching operations can pose.

Similar to river crossings, the training for breaching operations can be split into parts and brought together for final confirmation, with the desired outcome being easy to assess in that the battle group or brigade has either crossed the obstacle or not.

Reserves and Uncertainty

As with training for the defence, much discussion about the correct use of the reserve will be context-dependent on what circumstances the plan demands a reserve to address. As previously stated, once committed, reserves rarely return to a condition where they can "go again" within a useful time frame. Discussing the correct constitution and employment of reserves strikes the heart of how a formation or unit seeks to fight. "Simple plans and large reserves" are sound doctrine.

Reconnaissance only works if an immediately available reserve can exploit the conditions that the reconnaissance force discovers. The problem is that this places your force into an inherently dynamic condition where speed of reaction, better understanding and decentralised command will all have to come to the fore. Armies that can do this will be better trained than those which cannot, so it is a useful objective to aim for. Confidence in encounter battle will only come from extensive field training and, to some extent, the selection and training of commanders who are comfortable with the high levels of risk and uncertainty inherent to both and possibly unacceptable to policymakers because of the risk of high casualty outcomes. Much of this may seem a long way from the concept of the reserve, which is supposed to protect your force better against uncertainty because you can commit in support of those who find themselves in danger. That must be balanced against the clear benefits of reinforcing success rather than shoring up failure.

Much of this would rest on the higher commanders' overall intent and thus would be mission-specific. However, biasing tactical doctrine, as in "how to fight" towards unpredictable and chaotic conditions, demands a solid understanding of what the reserve is for and how best to use it. The bottom line might be that the size of the reserve and how it is used may be strongly correlated with the relevant commander's understanding of his mission and how he seeks to accomplish it. The simple question as to the intended role of the reserve or various units and detachments that might generate a reserve could well be used as a mirror to overall competence.

Conclusion

Hopefully, this chapter has shown that training for the offence is not merely about conducting choreographed battle group attacks dug in enemy sub-units. Like the defence, it is about understanding and thus mastering what you can train to do better than others in a way that makes its execution under the most demanding circumstances more likely to succeed.

Endnote

1 Ops Box Manta would qualify as a "Named Area of Interest" NAI in UK/NATO doctrine. An NAI is "where to look". Functionally an Ops Box is a bit of battlespace where an operation will be mounted – thus attached to a set of orders.

14

CLIMATES AND TERRAINS

This chapter might normally be called "warfare under special conditions." However, desert warfare is not special for many nations, including Israel. Israel also must contend with the highly complex terrain of South Lebanon, which a simple one-word description cannot define. For India, jungles and mountains are normal. Terrain and climate are vastly important in any domain of warfare, bar perhaps space. For an army that seeks any form of global or even regional relevance of expeditionary capability, the need to comfortably operate away from their native terrain and climatic conditions is non-discretionary. Even national defence forces must have a high degree of comfort across varying terrain and weather conditions. That has cost and training implications. It might also have equipment and force structure implications.

Terrain and climate are never homogeneous. There is flat jungle terrain and mountainous jungle, mountains in the desert, and mountains in the Arctic. No part of the planet remains "terra incognita" because it has not been mapped, visited or photographed from space. All of that information is now open and easily accessible. The question for anyone concerned about preparing an army for warfare under modern conditions is, "So what?"

As almost anyone who has ever carried a rifle and spoken on a radio knows, the effort to be a good soldier on a dry summer day with a cool breeze in the lush green rural landscape is nothing compared to that of a night of cold lashing rain on a rocky hillside with small scrubby bushes and no natural cover. What pains the soldier can well upset the whole force to varying degrees. The stark contrast in performance between Argentine and British soldiers in 1982 attested to just how far training and simple skills or personal administration can take you. More to the point, the structure of the British Task Force Land component had been optimised to operate under the conditions found on the Falkland Islands thanks to 3 Commando Brigades' extensive arctic warfare training.

From an Israeli perspective, it is a little-known fact that the Golan Heights has an annual rainfall like that of London. 90 per cent of that occurs between January and April.

The flow-down effect is that any Israeli vehicle must remain serviceable while parked in the open for three months of torrential rain and several nights of below-freezing temperatures. This isn't extraordinary, but a lesser trained force will have difficulties maintaining immediate readiness under such conditions.

In 2017, while conducting field trials at the British Army Training Unit in Suffield, Canada, I experienced the weather change from a barmy 23 degrees to a sub-zero, life-threatening blizzard in less than four days. It was a testament to British Army training that both trial units, the Household Cavalry Regiment and the Royal Tank Regiment, seemed to care little about how bad the weather got.

Terrain issues should be obvious but are often not adequately considered. Road network density is mostly a direct outcome of population and agriculture, all with climatic and terrain relevance. This is strongly related to issues such as the load classification of bridges and the repair state of the road network, as well as water and fuel infrastructure.

Therefore, discussing terrain in abstract terms like jungle, desert, arctic, and mountain makes little sense.

During my time in command post exercises in 1 and 3 UK Divisional HQs, I witnessed scenarios focussed on non-European terrains in countries described collectively as the Decisive Action Training Environment (DATE). This always seemed to generate useful insights about equipment and sustainment limitations, so much so that these were often glossed over to keep the exercise running. Regardless, the approach had considerable merit compared to assuming the baseline for modern peer-to-peer conflict was Poland or the Baltic states; thus, much of what informed this work was based on geographic locations for wargaming and military staff planning problems and estimates. These were:

- Sierra Leone
- Niger
- The Lake Baikal Region, Russia
- Northern Honshu Island, Japan.

The choices were partly informed by the fact that I have travelled in all these regions except the Lake Baikal Region; thus, I didn't have to rely on maps for what the ground actually looked like. The main reason for their selection over other options was that each consistently forced addressing manoeuvre and sustainment problems. Japan was the most surprising of all of these, and I learned much from working with a US Army Brigade HQ using this terrain for a planning exercise.

The overall aim here is to draw insights into training, equipment, and organisation from each region. These are unrelated to any defence planning scenario system, and all that follows will generally be agnostic of local politics, culture, or historical relevance.

Issues connected to so-called "urban operations" will be considered elsewhere.

Some readers may be inclined to assume that these geographic locations are axiomatically associated with a certain threat, such as Sierra Leone being nothing more than African warlord-type militias. There is a real danger here; if you assume the enemy to be incapable to the degree that nothing about them drives your tactical behaviour, then why would traversing the terrain and sustaining yourself in it be any challenge? As we shall see these can be dangerous assumptions.

Sierra Leone

This study's offence/defence scenario was an invasion towards the capital or major towns from neighbouring Guinea or Liberia, agnostic of Sierra Leone's current political or force structure realities. Several river systems, forest terrain, and low-lying marshland dominate Sierra Leone. There are five major towns. Two other rivers mark most of the northeastern and southern borders. The ability to cross rivers is therefore of some importance. Even in the country's north, these can range from 30 to 200m in width.[1] The banks are often heavily forested or on very soft ground to further complicate matters. Several rivers have extensive areas of rapids and midstream islands. There are a few tactical circumstances where river crossings were not a major consideration.

There are 935km of main roads and highways with all-weather road surfaces, but the road network is badly affected by the rainy season in all other cases. The larger towns outside the capital are generally crossroads or built around river crossings.

The countryside is mostly a patchwork of villages, pastures, arable land and jungle between the towns and rivers. The terrain offers excellent and effective natural concealment. Still, armoured or mechanised manoeuvres, while possible in some areas, will be severely constrained unless conducted by all-terrain vehicles able to swim and traverse the ground with less pressure than a human foot.

In terms of sustainment, ground lines of communications will be fixed to the all-weather surfaces during the rainy season, which any marginally competent enemy would find easy to monitor and interdict. During the dry season, there are more options, but not to the extent that it would create substantially more opportunities to manoeuvre. The climate also affects the flow rate in the rivers to the degree that it could affect the fuel usage rate of workboats, as is the case with some European rivers. This is only of concern if sustainment operations are restricted.

The significant factor of the rainy season (May-Oct), with between 300-670mm (14-26 inches) of rainfall a month, is it effectively negates useful UAS operation for most days, with cloud bases below 500m ASL not being uncommon. Heavy tropical rain severely impacts the performance of thermal imagery, whether vehicle-mounted, handheld or airborne. This type of weather also impacts all air operations. The use of helicopters is a significant enabler in terms of being able to mount and sustain all forms of operation in the Sierra Leone scenario. What became apparent in some of the operations studied was that in terms of peer opponents, the ability to deploy, sustain and recover dismounted MANPAD teams onto prominent terrain features significantly threatened and constrained the enemy's air movement and support plans as well as any UAS attempting to move below the cloud base. This was a course of action open to the enemy if they were to dedicate enough effort to doing the same, and it seems likely that such action would severely constrain air manoeuvre were the enemy to be proficient in using MANPADS.

Where UAS did seem to have potential was in the load-carrying variants described in Chapter 5. This made it possible to sustain remote MANPAD locations on high jungle-coated ridges only accessible on foot. It also enabled a dismounted screen spanning some 15km to be sustained under conditions that would have made it untenable without that capability. However, the viability of this type of operation at the height of the rainy season is unknown.

Excepting the climate difficulties, this scenario highlighted the use of heavier loitering munitions. This suggests a departure from normal concepts of operation as 155mm close support units may be hard, if not impossible, to sustain. This means that longer-ranged fires such as rockets and loitering munitions may be co-opted into close support roles, albeit limited to engaging point targets. This suggests a wider overall insight to be had here concerning how terrain and climate impact the force structure.

Regardless of, economy, history or politics, does most of the terrain in Sierra Leone prevent or severely constrain the operation of a force looking like a traditional combined arms formation? Armour saw extensive use in Vietnam and Southeast Asia and was deemed decisive in the Burma campaign of WW2. There are strong indicators that some armour and some artillery are useful enablers. Amphibious armoured All-Terrain vehicles appear critical to effective manoeuvre in most areas. Close air support would offer an advantage if target accurate targeting data were ascertainable.

Again, it is not the purpose of this section to predict what Combined Arms Force on Force warfare in Sierra Leone would look like. It is to ask what impacts the terrain would have on the force structures discussed so far. The answer appears very little unless the default setting is a Cold War-style armoured force. The previously proposed force design idea in Chapter 5 has infantry units equipped with amphibious armoured All-Terrain vehicles, an overall approach focusing on low weight, cost and complexity. Large forest and jungle areas effectively negate reliance on sensor-based UAS, as does the climate, so training in dismounted ground reconnaissance has primacy. In some respects, the expectation of what might impact the force structure seems to be cultural, as the jungle is known to force a high degree of emphasis on low-level infantry skills and minor tactics.

This is hard to reconcile with the need to dig in against 155mm airburst artillery landing on your position or how to sustain a Brigade along one muddy track. There is no reason not to suppose that two peer competitors will not engage each other in similar terrain. Assuming the jungle is the domain exclusive to third-world warlords and light-role infantry seems to be negligent.

Niger

This scenario relied on essentially the same construct as for Sierra Leone. Niger is the central Sahara, so it is hard to imagine anywhere with more stereotypical desert terrain. That said, the country spans terrain as diverse as the bleak and rugged Hoggar Mountains in the North to the lush fields around the 500m wide Niger River in the South. Going to the desert doesn't mean ditching the bridging kit. The Nile, Tigris, Euphrates and Suez Canal should be an obvious reminder.

The received wisdom of many is that ground manoeuvre in the desert is mostly unchecked by terrain. That may be true in certain areas, but it is the opposite in other areas because of boulder fields, wadis, dunes, and prominent topographic features offering multi-kilometre sightlines. In Niger, the areas between Bouza and Tahoua vary not much more than 100-150m in elevation. However, the frequency with which that occurs would prevent any armoured or mechanised unit from exploiting such terrain at any speed. Desert terrain usually presents few limitations of dismounted manoeuvre, bar high temperatures, which are less of a problem at night. Except for the flattest and most featureless areas, deserts can offer useful cover from direct fire and observation. Given that dismounted manoeuvre is almost silent, so under most conditions, a patrol with light scales should be able to move more than 20 km during one period of darkness, which in Niger is between 8 and 10 hours. While it mostly lacks vegetation, the desert surface is almost infinitely variable to the degree that maps contain little to no information about what to expect. Routes previously cleared on foot are of more value than many appreciate.

In terms of concealment, the desert ought to present an ideal environment to employ small UAS, except a daily average wind speed in February of 26 kph, gusting to 41 kph.

In July, that figure is 17 and 23kph, respectively. That is problematic for even a military UAS system capable of 129kph (70 knots). The Russian Lancet loitering munition has a reported cruising speed of only 100kph. Such wind speeds are substantially limiting for a civilian specification "quadcopter" capable of only 60-75 kph. However, these are not operating limitations for larger UAS operating from runways, as current US operations of MQ-9 in this exact area demonstrate. Surface temperatures and sand abrasion will present problems. These are issues that affect both manned rotary and fixed-wing aircraft, so they need to be considered in the wider conduct of operations. If your force relies on air observation, multi-day sandstorms might disrupt a surveillance plan to the point the enemy can exploit it.

Desert Mountains present several compound problems. Not only do they canalise any likely lines of approach or manoeuvre, but elevation and the lack of vegetation can maximise the effectiveness of screening forces. However, accessing the relevant observation points and being logistically sustained for any period is not simple, with water being the obvious limiting factor. Sustaining and manoeuvring the screening force is also canalised and predictable. The ability to scavenge sufficient water from the environment in terms of wells or natural cisterns in rock formations does appear to be a significant enabler for small, dismounted patrols.

The surprising insight as concerns Niger was that planning estimates revealed that a small force dispersed over a large area could be sustained under desert conditions, given adequate planning and forethought. This would seem to support the design for Light Cavalry type units adequately supplied with an anti-tank capability. Small vehicles can be easily hidden and dispersed outside any reasonable artillery footprint. Where large areas of unimpeded manoeuvre exist, the opportunities to attack and interdict ground lines of communication are obvious. The communications distances are vast, with a road move between Agadez and the capital Niamey being more than 900km, and even Niamey is 1,000km from the nearest seaport. From a planning perspective, the real insight was that while small-footprint sustainment was viable, it was mostly the only option, meaning only a small number of units could be supported and sustained across the distances inherent to the region.

Lake Baikal Region, Russia

This scenario assumed an attack or defence focussed between the Chinese border and the Western end of Lake Baikal. Assuming the reader can look at a map or satellite imagery, then the description of a large, forested area with a sparse but extensive road network connecting numerous towns and settlements distributed across large open river valleys with fields and marshland is hard to define in one word. The main area for study was Buryatia Oblast to the South of the lake. The complexity of the terrain should be placed secondary to the distances involved. The climate is one of short, extremely cold winters of down to −20F (−29C) but with paradoxically low snowfall. In contrast, the Irkutsk Oblast north of the lake sees an average low of −13f (−25C) with thick snow coverage. Daylight varies from 8-16 hours. The short summer sees temperatures rise to 25F (76C). It is a long way from the Arctic Circle, but all the arctic warfare skills will be required. As with Sierra Leone, this terrain emphasises amphibious and all-terrain armoured vehicles. Low temperatures present the standard problems for aircraft and UAS operations. Very few UAS have anti-icing equipment.

Deep snow, where it occurs, is the greatest limit to any manoeuvre unless using specialist tracked vehicles or snowshoes and skis for dismounted manoeuvre. In the absence of snow, there are large areas with sparse enough tree cover to allow vehicles to move between tracks and roads. There are also thickly forested areas impassable except on foot. Where roads are easily blocked due to the channelling inherent to valleys as consistent, the useful ability seems to be that of dismounted infantry to infiltrate past detected positions if they can move across 25-30km of rough terrain to conduct an attack on the rear of the blocking position or block reinforcement.

Rivers and swamp areas also impact manoeuvre and mostly occur within open areas of arable land in wide open valleys with extensive grassland areas. This produces a near-binary effect of forested hilly areas immediately next and flat open areas with little intermediate terrain types.

The same condition that made MANPADS disproportionately effective in countering air manoeuvres in Sierra Leone also seems to exist in this region, except in most cases, they are limited to point defence given the vast area size. The tyranny of distance, thus sustainment and dispersion, strikes the understanding of what might be considered a Divisional, Corps or Theatre asset. In a traditional system of battle space frontages and depths, it is comparatively easy to understand the need for a close support regiment within 15-25km of the formation it supports. Once sensors and communications enable greater detection ranges, the apparent need is for any sensor to be able to task any weapon. This condition is entirely forced by the lack of a road network and extended ground lines of communication. As with Sierra Leone, systems normally associated with the deep battle are forced into close support because close support weights of fire are not logistically sustainable at reach, nor are heavily armoured platforms.

Northern Honshu Island, Japan

This terrain area assumed operation moving on a North-South Axis from the northern edge of Sendai to Aomori on the northern tip.

The key characteristic of this terrain was an extensive road network and large, dense urban areas in wide, flat valleys separated by thickly forested mountains. The climate is warm and wet in summer and cold, with heavy snowfall in winter.

The high-density road network is mostly concentrated in the flat valley areas but also provides more options for crossing the mountain areas than would be the case in a less developed country. The valley areas are generally urban or rice farming areas with large rivers and drainage and irrigation ditch networks. That said, the vast proportion of the terrain is thickly forested mountains, forcing manoeuvre and sustainment into the valley areas. Movement through the mountains is likely easy to block or interdict, but the number of small roads and tracks, even within the forests and mountains, means this would require more effort than would be the case in less developed areas.

Perhaps the most useful insight is the structure of the JSDF ground forces' understanding of how their nation's climate and terrain have shaped their equipment choices. Their main battle tanks are significantly lighter than their Western counterparts, with the Type 90 being some 50 tonnes and the newer Type 10 being between 40 and 48 tonnes, depending on load and configuration. Regarding bridge classifications, 65-70 tonne vehicles would be severely constrained. Exploiting narrow and winding mountain roads would also seem critical, albeit less hard to ascertain regarding their published force structures.

Northern Honshu seems to confirm the clear conclusion that mechanised manoeuvres have to be able to traverse difficult terrain, and dismounted manoeuvres have to be able to move large distances through rugged terrain impassable to vehicles.

Conclusion

This section confirms that forces configured within cost, weight and complexity constraints seem to have more options than ones that are not. The traditional approach to jungles and mountains of defaulting to helicopters may not always be feasible, given any competent enemy prepared to contest your use of the air.

Three major insights seem apparent.

First, given large distances and/or a restricted road network, a unit's ability to sustain itself without resupply seems important. How long that may be is unknown, and it needs to be balanced against not impeding manoeuvre with large amounts of "on wheels" stores. That said, it must also be realised that the constraint that sustainment applies acts as a useful counterweight to the force becoming over-extended and unable to concentrate or conduct a regain if necessary.

Second is that amphibious capability seems to buy more than it costs to give commanders options, as does some form of all-terrain APC and load-carrying capability.

That has to be balanced against the other clear need to be able to move long distances across sparsely populated areas.

Third is the truncation of Deep fires into the close support roles because, under some conditions, it will be challenging to sustain close support gun units.

It would be perfectly true to say that all the constraints the terrain forces upon us also force upon the enemy, but it could be that the enemy has had more time to prepare and ground dump large stocks of ammunition and stores.

This chapter should have asked what additional training, equipment, and organisation a force needs to operate in any of the scenarios examined. A clear understanding should emerge from asking these questions.

Endnote

1 The Rokel River exceeds 200m upstream of the Bumbuna Dam.

15

URBAN OPERATIONS

This chapter is not so much about the supposed complexity and difficulty of urban operations as it is an examination of the popular mythology that has given urban operations such prominence and the consequences of that for equipment, training, and organisation[1].

Urban operations are far less challenging than popularly advertised. The data is proof of that. Training for urban operations is comparatively simple and cheap compared to the training required for operating in the jungle or arctic.

A key point is that not every terrain with some buildings is "urban." Villages where every building can be engaged from outside its confines or large areas of dispersed dwellings interspersed with small agricultural holdings, as found throughout Russia and the Middle East, are not urban areas by any stretch. The concept of urban should be limited to large areas of multi-storey ballistically tolerant structures.

Work done by the UK's Defence Evaluation and Research Agency (DERA) in the 1980s showed that urban terrain was not a defender's paradise, nor was it inordinately difficult for the attacker. The very opposite was true. The attackers almost always won, and the defenders suffered high casualties in almost all cases for which detailed data were available. The defence has fewer advantages in the urban because the enemy can gain proximity without detection, and truncated sight lines mean it is easy to find the flanks and rears of enemy positions and bypass them altogether.

Notably, the deciding factor in Urban Operations was good training and supporting fires from AFVs.[2] Based on comparing historical analysis with instrumented troops trials using the Berlin Brigade, further research confirmed that urban operations usually, but not always, ended badly for the defender for very easily understood reasons. Skilled urban defences were rare but required pre-planned counterattacks best supported by armour. Recent experience of urban operations has seen the defender lose in almost every case.

Thus the idea that urban terrain favours the defender is wrong despite assertions directly contradicting that.[3]

Yet more evidence was apparent from work done by Christopher Lawrence of the Dupuy Institute outlined in his 2017 work, *War by Numbers*. This essentially confirmed the DERA findings. Using mainly WW2 data and summarised in a chapter called "Urban Legends,"[4] his additional findings showed little exceptional about the urban. Urban combat was not more lethal than non-urban. Nor was the claim that urban combat was more intense than non-urban supported by the evidence. Notably, the work done by the Dupuy Institute was commissioned in 2001 by the US Department of Defence and the data largely disproved urban as an exceptionally hostile terrain. This has been potentially in full view of the advocates of urban operations for a very long time but clearly and most likely deliberately ignored.

Even cursory analysis of commonly available data tends to support the above. For example, the Battle of Marawi saw 150 days of fighting where the defenders lost catastrophically, suffering a Killed in Action (KIA) loss rate of 6.52 per day compared to the attacker's 1.12 KIA per day.

Fallujah in 2004 was fought over 50 days and saw the US attacker suffer 112 KIA at a rate of 2.24 per day, while the defender suffered an estimated loss rate of 40 KIA per day.

Operation Protective Edge in 2014 saw 49 days of fighting, with the IDF losing 67 KIA, so a casualty rate higher than Marawi at 1.3 KIA per day, but only very marginally.

As of August 2024, the IDF had 329 soldiers killed, and 2,199 were wounded in 283 days of combat operations in Gaza so 1.16 KIA per day.

In contrast to all this, the Blackhawk Down event of 3-4 October 1993 saw 19 US KIA and 70 wounded, but that is an anomaly. On the 20th of June 1967, the British Army suffered 22 KIA and 31 wounded in one day of fighting in the city of Aden. Today, this incident is almost entirely forgotten, and almost no urban warfare literature mentions it.

This was also an anomaly. Neither event was indicative of a trend, nor did they occur during warfighting operations. In 1965, the US Army suffered 237 KIA in 5 days in the Ia Drang Valley, far more than the oft-cited 72 KIA across eight days in the fight for "Hamburger Hill."

Additionally, the often-cited defeat of Russian Forces in Grozny offers no more caution than what a lack of training and poor planning means for any soldiers. It is likewise notable that those urban operations with definably high casualty rates are almost exclusively Russian or Soviet.

Most urban loss rates do not even come close to the non-urban casualty rates, such as the average of 130-150 KIA dead per day for non-urban battles as diverse as El Alamein, Iwo Jima, Okinawa, or the entire Israeli KIA for the 1973 War.

The often cited examples of Stalingrad in 1942 and Berlin in 1945 are often unaware of the fact that in Stalingrad, most of the casualties occurred outside of the city itself. The same is not true for Berlin, but large amounts of casualties were suffered outside the city, further inflating the number. The argument that poor Soviet command decisions were the real cause of high losses does carry some weight. Regardless of that, Berlin 1945 is an anomaly, not a norm.

As of the time of writing, no open-source data set suggests that urban combat in Ukraine has been specifically more costly than rural combat. Even if there were to be one, it is of doubtful relevance given the Russian levels of training.

Little statistical evidence suggests that urban areas are more lethal or dangerous.

Most of the evidence appears to be anecdotal. The fact that sometimes high casualties occur in urban combat does not make it exceptionally difficult to operate in.

Moreover, why would an environment comprising overwhelmingly ballistically tolerant structures, thus cover from every sort of fire, be more dangerous than woods, fields and hedges, which offer far less cover overall?

Thus, it appears that what some armies believe about urban operations is not based on actual operational analysis but on narrative literature designed to promote agendas and reputations. The literature falls into two distinct categories. The first is the historical narrative, which emphasizes the intensity and difficulties of urban combat. The second is a predictive approach that urges greater study of urban operations because of their increasing prevalence and difficulty, which is contradicted by the historical data.

Thus, the historical narrative lacks an evidential basis, and the warning for the future dates back well over 30 years, yet the prophecy has not come to pass. Worse, there is incredibly little written about urban warfare training. Most of what is written is based on the "difficulty narrative". There is now an entire body of literature seeking to explain why urban operations are the future of warfare, but with little or no evidential basis to explain why other than to say the world is becoming more urbanised.

Of course, urban operations are difficult, but the problem with "urban exceptionalism" is the defining idea that they are more difficult than anything else. For example, in 24 hours, the US 1st Infantry Division lost 300 men killed between 29th and 30th of November 1944 in the Hurtgen Forest. Likewise, the previously mentioned Ia Drang Valley. Is fighting in forests easier than fighting in urban? The amount of writing about the nature of fighting in forests and jungles is minuscule compared to that of urban.

Yet, it produces high casualty events and requires substantial skill. Nothing in the research for writing this book suggested the need for a chapter on Forest combat. 30 percent of the Earth's land is forested. Only three percent is urban.

So What?

The myth that urban warfare is uniquely demanding is a fallacy based on a body of narrative literature, not operational analysis. It gets written about a lot because of how it is perceived, not how history shows it is. Much urban operations literature is arguably irrelevant. Much military literature seems to have little effect on what real soldiers do. The literature mostly comes from academics, Think Tanks, or military historians with books to sell. The problems occur when armies believe data-free propositions urban operations are uniquely hard because it speaks to how they understand the evidence. That is a problem because the story has become the driver, not the evidence. It is akin to institutional blindness.

This can manifest in such ideas as armoured vehicles specifically designed for "urban combat" or co-opting ideas such as "hardened logistics" as an urban operations requirement. Both examples rely on their technical merits being exclusive of the urban operations. If an AFV specifically designed for urban is more useful in all terrains than one for non-urban operations, then there is little to discuss. The problem occurs when the opposite is true. If urban is deemed exceptional, it will demand heavier, more expensive, and more complicated technical approaches. The fact that the Dupuy Institute notes that, by and large, urban operations do not lead to higher AFV loss rates is evidence best ignored by those promoting the idea.[5] Given the amount of cover available this should not seem surprising.

There are some important things to understand about urban operations. The first is to understand why the defender logically loses, should he seek to defend from strong points. Defending any structure in the urban is predicated on an all-around defence and the need to hold to the bottom two floors of any structure (plus any underground access).

If the enemy manages to overrun the ground floors everyone else in the structure is isolated and at the mercy of demolition or setting fires. Holding ground from within buildings absorbs huge amounts of dismounted manpower, so "strong points" are not the way to go. A strong point is eventually a street address or 10-figure grid which is then just a target to be isolated and reduced unless it can enable wider blocking actions or counter-strokes by large reserves. Add to that the fact that limited fields of view mean mutually supporting direct fire is hard to facilitate. Your urban defence has shrunk to just the nine large buildings your nine rifle platoons in your battalion can hold and probably less if you want a mobile reserve, which will be essential. The number of buildings you can hold maybe even less certain by the scale of anti-armour weapons. An infantry brigade of four battalions with nine platoons can only hold 30-70 large buildings. Those 30-70 large buildings must contain all the logistics needed to sustain the defenders because re-supply and casualty evacuation is unlikely once fixed inside the building.

Those 30-70 large buildings must somehow not be identified or surrounded so that the defenders are fixed inside and then buildings reduced by indirect fire. That is very hard to achieve, so the defender almost always loses unless well dispersed into a mobile defence cued by ISTAR and controlled by superior C2 – as is the case in the non-urban.

The urban defence relies on ambushes, counterattacks from flanks and rears, and the need not to become fixed.

That leads to the second reason that, given the first, the fundamentals of modern defence apply just as well in the urban as in the rural or any terrain. If you cannot conduct or train for a coherent and sustainable defence in the non-urban, you cannot do the same in the urban. In fact, given the oft mentioned channelisation of enemy approaches, it should be easier to conduct a mobile defence in urban than rural terrain where the manoeuvre is less constrained. Yet this touches on a paradox that a mobile defence of sufficiently large urban terrain could be more effective than in the rural. That effectiveness is born of mobility and manoeuvre, but these advantages lie with both the attackers and the defenders. Thus, the advantage lies not with the attack or the defence but with the more skilled manoeuvre and, thus, the better C2. The absence of these two simple points from much current expert opinion should indicate that while much has been written, some simple, practical aspects have not been considered despite being well-known in the 1980s and 1990s. The C2 issue is especially interesting as the Israeli Defence Force implicitly describes urban operations as more a C2 problem than anything else.[6]

My experience of the IDF is that they view command and control as more problematic in the urban than manoeuvre or sustainment.

As has been constantly discussed in this work thus far, military thought is riven with the fallacious foretelling of future war and warfare, yet basics endure. If you trained for urban operations in the 1980s or 1990s, you have little to make up for being well-trained today. There has never been a time in recorded history when fighting in towns, cities, or villages has not been prevalent. Likewise, sieges endured, be they Leningrad, Dien Bien Phu, Khe Sahn, or Sarajevo – notably, two were airfields, not cities. It seems fair to suggest that "sieges" will be as common as they always were.

Good Urban Training

Urban training is logically constrained by time and budget. That said urban training facilities need not be expensive or extensive, despite the fact both expensive and extensive facilities exist. The Israel Urban Warfare Training Centre at T'selim contains more than 450-470 buildings and structures within 0.18 square kilometres of terrain. This is exclusive of a dedicated urban live firing complex built close by. Europe's largest urban training facility is the French CENZUB or "Centre d'entrainement aux actions en zone urbaine" between Reims and Saint Quentin in Northeastern France.

Urban operations essentially consist of three basic requirements:

- Attacking or defending from one structure to another
- Attacking or defending within a structure
- Manoeuvring between and within structures.

Nothing about that is particularly complicated. Structures can be buildings, complexes, underground facilities or tunnels. It does not matter.

Since urban terrain substantially degrades stand-off sensors, manned reconnaissance is needed, which is usefully supported by unmanned systems. For infantry, the initial primary training objective is dismounted reconnaissance and observation post-training. The counter-surveillance training given to infantry for non-urban environments needs very little additional training for urban environments.

Navigation in urban environments is tremendously simple compared to the jungle, forest, or desert, and fixing the location of a target is far less ambiguous. These are already substantial training advantages. Initial approaches should focus on locating the enemy.

Hence, the overall approach to urban operations is coherent with the Find, Fix, Strike and Exploit Core Functions framework for all environments. How to defeat an enemy force within a building is strongly related to finding them in the first place and fixing them within the structure so that a higher enough level of casualties can be inflicted to force them to surrender. Fixing ideally means they have no escape, and counterattacks by adjacent elements can be defeated. The main problem with many historic approaches to urban training is that they have focussed on fighting within the building the enemy is contesting rather than reducing the enemy by fire from outside it. Regarding large or multi-story structures, occupation or denial of the ground floor usually decides the fate of anyone higher up. Modern infantry have access to large quantities of projected high explosives compared to WW2. For example, Soviet infantry in Berlin in 1945 were entirely reliant on sub-machine guns, grenades and direct fire from tanks. A modern infantry platoon has substantially more options in the shape of 40mm grenades, 66mm anti-structure munitions, 84mm high explosive rounds, and many similar capabilities.

The presence of civilians will restrict the use of force that can be applied in any given situation, but context is critical. Each nation will have specific rules of engagement, and the Law of Armed Conflict (LOAC) will most likely supply non-discretionary conditions, but context is critical. The main concern of most commanders will be the issue of proportionality in that the harm inflicted on any civilians, or their property, must not be more than that required to the military objective.[7] This is extremely context-specific and essentially down to the judgement of local commanders. However, this highlights the difference between warfighting and security operations in terms of why civilians would be present in an area during warfighting operations. If an enemy force decides to launch operations from an area with civilians present or defend the ground without evacuating civilians in the hope that the presence of civilians will act to their advantage, that is contrary to LOAC. Regardless, training needs to account for this. Much of the academic fascination and faux-urban expertise is a direct outcome of the last two decades and the idea that warfighting operations would be conducted with a static civilian population.

The Israeli experience is that in Southern Lebanon, the population will move, and in the Gaza Strip, the population can be told where to move and will mostly comply. The methodologies for accomplishing such compliance will be specific to context and circumstances, so they fall outside the scope of this work. Still, all competent armies should have developed options for making this happen.

Above the individual and small unit tactics for operating within the urban environment, the next level is cooperating with supporting arms such as armour and artillery. This is comparatively simple regarding tactics, techniques and procedures and can be effectively progressed using simulation and even "ISO container villages" to practice cooperating with armoured vehicles.

Physical training doesn't need to be above sub-unit, bar-specific training packages for unit echelons, such as logistic support considerations. A unit or battalion commander with three or four well-trained sub-units is as well prepared for operations as physical training will allow. The actual mechanics of manoeuvring forces in the urban terrain are best done as a planning exercise, staff ride, command post exercise or terrain study. Few nations will tolerate urban training in real cities and towns.

Generally, any combat unit must probably complete 5-10 days of specific urban training every year. Command staff and planners should be about the same. This is training specific to urban because all training should contain some elements tangentially relevant to urban operations, as most operating environments contain some degree of human habitation.

Being competent and prepared for urban warfare is not hard or complicated. It just requires the same level of attention that might be focused on any operating environment.

The restrictions on force inherent to a civilian population need prior thought, but they cannot and must not restrict the military instrument to irrelevance to the enemy's advantage. That is not the reality of war, nor was it ever. Overwhelmingly, the intent to use armed force against only armed forces in line with precision, proportion and discrimination will seldom render the actor at fault.

Conclusion

The remedy to the myth of urban operations is urban operations training. Good urban operations training gives soldiers accurate expectations of what is required and gives staff and commanders a sound understanding of what is possible and what is not. If an army is reluctant to operate in urban terrain, then something has gone badly wrong with its leadership. Likewise, an army that has focussed on urban operations to the degree that it is less than competent when presented with rural or complex terrain has a similar problem. There has been no time in the last hundred years where this observation did not make sense. Beware of those who say otherwise.

Endnotes

1 I worked as a contractor on the British Army's Urban Doctrine rewrite in 2014 and also worked with the Royal Marine Urban experimentation company in 2015. Over a period of 12 years, I have also observed at close proximity IDF Battalion, Company and live fire urban training. The urban live firing included MBTs firing live rounds in close support.
2 Rowlands, David, "The Stress of Battle," The Stationary Office 2006 Page 90.
3 The entire underlying assumption of the British Army's "Urban Warrior" experimentation program was based on this idea.
4 Lawrence, Christopher, *War by Numbers*, Potomac Books 2017
5 Ibid.
6 Personal observation based on talking to IDF officers and observing training at the Urban Warfare School over the period of a decade.
7 This should NOT be taken as legal opinion.

16

AIR AND LITTORAL

Air and Littoral operations merit a degree of specific study for two reasons. First, we must ask if units dedicated to such operations demand specific organisation, equipment and training. The second is that much about the traditional models for such operations have evolved considerably in recent decades.

Air and Littoral could have been expressed as parachuting, airmobile and amphibious, but each term comes with considerable social and cultural baggage in almost every developed army. This impediment almost always retards useful insight. What follows here is not a debate about the merits of each or any, but more a discussion as to the "so what" of such capabilities.

Before proceeding, we need to clarify some terms.

Parachute Operations are those where troops and equipment are delivered into theatre or operations using parachute equipment.

Air Manoeuvres/Mobile Operations involve troops using helicopters to gain or sustain an advantage over the enemy.

Air Landing Operations is the use of fixed-wing aircraft to position troops.

Littoral Operations are those where troops use the sea to manoeuvre onto and off the land to gain or sustain an advantage over the enemy.

More complex explanations may be better suited to doctrinal arguments, but these will suffice to boil down the emotional reactions and gatekeeping. The level of command involved is key to understanding the context of all these operations. If the operation is part of the theatre plan, then it will get access to theatre assets. If part of a Divisional plan, then the assets allocated may be substantially less.

Military Parachuting

Parachuting is a specialist skill requiring volunteers, and even then, not everyone can do it to the degree that they will, without question, throw themselves out of the door on a dark night, in close proximity to the ground, with only a few seconds to make life-or-death decisions. In this regard, paratroopers are distinct from others, but beyond that, lots of the differentiation is merely cultural. Like sniping, parachuting is seen as a status qualification. So much so that attending the basic parachute course and qualifying to wear parachute wings is used as a reward in some armies. Almost every special forces unit baselines its capability to parachute insertion, even when that event is deemed extremely unlikely. Yet there is far more to a parachute capability than being able to drop a platoon or sub-unit out of an aircraft. There is more to parachute capability than just parachuting.

It must be conceded that even outside the domain of special forces, having parachute-trained infantry and even some supporting arms is something developed armies should seek to maintain. The simple reason is that you never know when it may be required. The problem arises when those who cannot be objective about parachute capability, for whatever reason, claim that it is a thing of the past or cite unknowable future circumstances that would make it a poor investment or use of funding.

Excluding special forces, the key point to realise is that parachute operations are rarely, if ever, decisive in and of themselves. Parachuting is a means of delivery that allows the placement of forces onto terrain that enables other forces to complete their mission. The seizure of bridges is the most obvious, as in Operation Market Garden, the failure of which is largely attributable to the 82nd Airborne's failure to seize the bridge at Nijmegen.[1]

If a parachute force cannot seize and hold key terrain or infrastructure, it is of substantially less value than one that can. Parachute operations must plan to terminate when other ground forces extract or relieve the air dropped component. Accepting this premise is key to developing and resourcing a modern parachute capability.

Parachute operations can occur beyond the range of helicopters and usually deliver more mass over a shorter time. For example, a CH-47 Chinook can move 40 troops over a 160km radius of action.[2] With a low-level flight profile, a C-130J can deliver 64 paratroopers to 1,000km or more.[3] Depending on the planning metric, the Drop Zone for a single pass to deploy 64 paratroopers (2 Sticks of 32) at a low level would be at least 500m x 2,500m. Thus, 12 platoons of 24 infantry would take just over 8 Chinooks or 5 C-130J with some capacity of Battalion and Coy HQs for a total of 320 in each case. C-130J is merely used for familiarity. C-17, KC-390 and A400M are obvious alternatives, but they would require fewer aircraft and larger drop zones. The most apparent circumstance where parachuting maybe employed at ranges helicopters could transit would be to drop troops behind air defences or from an unguarded and thus unexpected direction.

The main limiting factor for any significant parachute employment is the Drop Zone's size, which depends on a large area of flat, unforested ground around, on, or close to the objective. If the objective is an airfield, then this is obvious. However, this may not be possible if the objective is a bridge in a heavily forested or jungle area or mountain pass. Smaller Drop Zones can be used, but aircraft must make multiple passes.

The alternative is to drop from an altitude of 10-12,500ft using static line deployed square steerable parachutes and have the troops navigate to bring themselves in on 400m x 400m drop zone.[4] The training debt would mainly be learning to manage a steerable canopy and calculating the deployment of the sticks in terms of wind. Doing this at night would also present additional training, especially if using Night Vision goggles to find the drop zone. Higher altitudes require oxygen, substantially more training and preparation, and expense. Free fall would be prohibitively expensive to train, so the option would still be for static lines, albeit at high altitudes and openings. High altitude means between 24,000 ft and 32,000 ft. Thus, depending on conditions, the aircraft can deploy troops up to 20-30km from the drop zone. 40km may even be possible with high-performance canopies. However, this does expose soldiers to extended periods in cold temperatures and on oxygen, which, as previously mentioned, may be seen as an excessive training burden. The expense must be balanced against the increased capability that long-range stand-off might present. Most debates about modern parachute operations hinge on the ability to conduct them with any amount of enemy air defence being present. Even poorly equipped forces might possess MANPADs, which can threaten aircraft above 15,000ft, albeit detecting and engaging aircraft at that altitude is not a given.[5]

Wider discussion and wargaming in the preparation of this work concluded that the ability to compress the drop zone via a static line with steerable canopies, jumping from below 12,000 ft, was probably the key enabler to the conduct of modern parachute operations. By far, the most prominent item in the discussion was the political permission to mount such an operation with any form of air defence threat without an existential risk to national sovereignty. In 2014, I was part of a Command Staff that conducted such an operation as part of Divisional CPX, which included 4 NATO nations. We planned and executed a low-level parachute drop to secure bridges across a canal at the foot of a large dam. While notionally successful, the umpires allocated such a high level of casualties to the operation that the Parliament of the nation, whose parachute unit conducted the operation, withdrew them from further combat operations. While anecdotes should not be taken as evidence, this clearly emphasises that high-risk tactical concepts need to be addressed within the context of the political reality that rules over all else.

Stand-off jumps are not limited to infantry. Large RAM Air canopies can be remotely guided into drop zones with loads up to 4,000kg, though loads substantially less than that are more likely so dependent on the number of aircraft and the tactical situation that it is perfectly possible to emplace an effective light infantry force with sufficient support weapons, sustainment and even a degree of mobility.[6]

Concerns about dropping stores, vehicles, and supplies should be understood as concerns about altitude and training being much less limited, so greater stand-off ranges can be planned for with few limitations. This also supports the re-supply and sustainment of non-parachute forces by parachute, which should be something most developed armies seek to have as some degree of capability.

Air Manoeuvre Operations

Air Manoeuvre is no different from the decades-old concept of air mobility. Using helicopters to gain and sustain an advantage over the enemy should not be overly complicated in training, command or doctrine. It should be as simple as possible and not require dedicated formations to execute. Some degree of specialist planning and coordination is required but nothing that cannot be imparted by a short period of specialist training and should not require dedicated personnel at either the Division or Formation level. Some people will baulk at this observation, but how much extra staff planning data and knowledge does it take to plan and execute a helicopter move, even when uplifting from several different locations to set down at several more, even at night and in bad weather? Deconfliction of air movement will be a factor, but not one that needs ballet levels of choreography. If the plan isn't simple, it simply won't work. This may include troops being dropped into a location in immediate proximity to an area struck moments before by either artillery or air interdiction. However, being comfortable with that level of risk is inherent to all warfighting operations.

All units with dismounted roles must be trained in helicopter moves and sustainment. This means junior commanders can talk aircraft into their location and judge that location as suitable for the aircraft to set down and then take off again, possibly close to the load, temperature and altitude limitations of the type concerned. This is unlikely with Chinook but very likely with some older or smaller airframes in terrain, such as the Hoggar Mountains.

The most dynamic form of air manoeuvre is an anti-armour screen's deployment, sustainment and recovery. This will require a dedicated and prepared force for something formed at the Corps or Division level. If no enemy armour threat is present, this will most aptly be described as the Air Mobile Reserve. Rapidly flying forward, small dismounted anti-armour teams under considerable uncertainty are only possible with lots of real-world training, where decisions made by aircraft and junior commanders are the judgments that prove decisive. Not only could this be anti-armour teams, but Forward observers and anti-air teams as well. Deconfliction between your anti-air and any close air support or attack aviation would need to be via battlespace and robust communications regarding when to switch between "weapons tight" and "weapons hold" for air defence teams. The condition of "weapons-free" is extremely unlikely in modern operations. This is demanding in terms of training but not impossible.

Modern navigation and digital data communications are substantial enablers compared to what was possible in the 1980s. For example, there is no reason why UAS cannot be used to allow commanders to dynamically task teams to be dropped onto avenues of approach it can identify and co-ordinate the screen with the conduct of the deep battle. The key point here is that using an airmobile reserve would most likely become part of the deep battle since it would be conducted beyond the range of the close support artillery. This will certainly be the case if co-ordinating with Loitering Munitions, Attack Aviation and Fast Air.

As with all aspects of modern operations discussed so far, the key to air manoeuvre of whichever type is training, especially if screening forces or reserves are going to be brought into action by helicopters.

Air Landing Operations

Transporting troops from airfield to airfield hardly seems to require any discussion except when it does. The most obvious exception is when parachute forces or others have seized the destination airfield as part of a coup-de-main, and the location needs rapidly reinforcing. The less obvious is the use of improvised operating surfaces such as highways, beaches or temporary runways prepared by engineers. Ideally, these will be suitable for traditional large transport aircraft such as C-130, KC-390, A400M and C-17.

The Israeli Air Force maintains the specialist "Front Landing Unit" ('Yahaq' in Hebrew abbreviation) to locate, survey, test, mark, and clear improvised operating strips suitable for C-130 operation. These sites can then be enhanced with portable lighting systems and instrument landing equipment, and a small, dismounted team provides a degree of austere air traffic control.

Most infantry units can accomplish air landing operations as a distinct tactical method as part of an airfield seizure with no specialist training bar the loading plans for the aircraft concerned. This will almost always be accomplished using traditional tactical transports with ramp loading systems.

Sustaining remote locations or detachments via routine military transport aircraft operations is well understood and practised, as is parachute re-supply. Air landing operations using light aircraft onto improvised or austere operating surfaces are not well recognised within many militaries. The training aim is to teach both combat and engineer units to identify or construct operating surfaces that will allow the operation of light aircraft or, in the immediate future, load-carrying UAS such as the Windracers ULTRA able to carry 200kg of stores within a 700-litre cargo bay out to mission radius of 500km.

This requires an operating surface of 150m. Assuming a cruising speed of 135kph means an 8-hour round trip equating to 600kg of stores per airframe in 24 hours. Of note, the latest CH-47F has a 370km loaded radius of action. The UH-60L can operate to range of 590km with a payload of 1,400kg internally. The current production de Havilland Canada Twin Otter (Series 400) can deliver 3,000kg of stores to a 600km radius but requires 350-400m of operating surface. Assuming warfare is about making the best use and allocating resources, preserving helicopter operating hours and costs should be more widely considered if operationally robust and proven alternatives exist, especially in protracted conflicts. The use of UAS should be considered as a baseline in this regard.

Littoral Operations

Any current discussion of littoral or amphibious operations should not be seen in the context of the current United States Marine Corps Force Design 2030 and its plethora of debates. Correctly employed, Marines and naval infantry are the land component of the fleet, supporting naval actions, so they are not part of the land force.

Land Force's littoral capability is using the sea to manoeuvre against a land-based enemy. While in some cases functionally identical, there are notable distinctions in that the Naval land components will use terrain to threaten enemy naval assets to seize or protect harbours and anchorages for the fleet. That is distinct from how the Land littoral component will be employed regarding training, equipment and organisation. The distinction described is a "platonic ideal", never likely to be a reality given budgets, service organisation and rivalry.

Littoral manoeuvre is most likely to take place within the context of theatre operations and to support or be supported by other land and air operations. The "so what" implications for unit organisation, equipment and training is that your force has to be deployable from sea-going vessels without the need to re-organise, re-train and re-equip. Much of this depends on the nature of the vessels concerned and the investment made by the Navy to support such a concept. Absent of that there must be a plan to use ships taken up from trade to do the same.

The Falklands War was won entirely by light infantry units usefully supported by artillery and one sub-unit of tracked reconnaissance vehicles. If unopposed, a modern light infantry unit should be relatively easy to move from ship to shore. The likelihood of an unopposed landing is greatly increased if the force can land at a time and place of its choosing and one not obvious to the enemy. This means a shoreline is less than ideal for landing vehicles or men. Unopposed beach landings are the historical norm compared to the iconic images of Dieppe, Normandy or Tarawa.

Helicopters are the most obvious choice; we have already discussed that in detail.

The major difference in operating from ships is that limited deck space can limit the speed at which troops can board aircraft, thus slowing the total time to move the force.

Apart from this, all other aspects are broadly the same. The limited deck space becomes more acute when recovering the force back to the ship, possibly under less-than-ideal conditions.

Moving troops with light scales ashore via boats such as Rigid Hull Inflatable Boats (RIBS), which can move 10-16 troops >25nm offshore, allows access to most shores short of sea cliffs. Shipboard crane systems can deploy most RIBS. Getting troops from the ship into the RIBS can usually be accomplished easily on calm seas, but it still demands training and familiarity. The issue of getting anything more than man-portable weapon systems ashore is where the complexity and cost start to creep in.

Given deep battle-type support, the infantry force can fight effectively. Loitering munitions and UAS can easily be deployed from ships. Still, the gun and rocket systems within the division would be mostly impossible to operate, even given the deck space to allow it.

The infantry force should be able to detect and direct strikes against any threatening enemy force and defend itself with guided weapons and medium mortars. This should allow the shipping to close with the shore and allow vehicles to offload. However, this will mostly demand either roll-on roll-off ferries able to approach the beach or jetty that enables some type of interface with the shore or use amphibious vehicles as covered in the infantry discussion. As for an interface between ship and shore, there is no reason why specialist pontoon systems cannot be used, as routinely employed by the British Army in the shape of the Mexefloat system. Beyond this point, much of what may or may not be possible depends on specialist military equipment, such as various landing craft and rafting systems. Critical to any sustainment ashore is the ability to get fully laden trucks from the shipping onto the shore and transfer stores and equipment at a rate that can support the forces in contact.

In terms of my own research, wargaming and planning exercises used to examine this form of operation not surprisingly concluded that the movement of logistics and casualty evacuation were the critical factors. So much so that operations were often impossible without helicopter support or load-carrying UAS, excepting a considerable appetite for risk.

In terms of cost, weight and complexity, the burden for the investment to enable littoral manoeuvre falls almost entirely outside the scope of equipment this work has considered. Adequately resourced with amphibious shipping, landing craft and support helicopters there seems little about this form of operation that is not either well understood or could not be addressed by rigorous training. Getting the shipping into the proximity to the shore to allow for such operations is a Navy or Air Force problem.

Endnotes

1 Poulson, RG, *Lost at Nijmegen* 2011.
2 UK Staff Officers Handbook 2001.
3 UK Staff Officers Handbook 2014 (Draft).
4 Based on discussions with a senior UK Parachute Regiment Officer with extensive military square canopy operation.
5 9K333 SA-25 Verba.
6 The US Army has been examining this for well over a decade in the shape of the JPADs project.

17

SECURITY AND COUNTER-TERRORIST OPERATIONS

Recent decades have seen much debate about "Counterinsurgency", with a proliferation of opinions and definitions. The view progressed here is that modern armies need to be able to target and defeat any element within a nation or jurisdiction that threatens Government authority and, by extension, the civilian population. It is about detaining or killing armed criminal elements who place civilians in danger. It is arguably what most modern armies are most likely to do daily.

The overall description of Counter Terrorist Operations is more constructive than "counter-insurgency" since terrorism can simply be defined as a criminal act within a jurisdiction. One man's terrorist is no more another man's "freedom fighter" any more than one man's burglar is another man's Robin Hood. Terrorism is a legally definable offence. This description can usefully extend to poaching gangs, narcotics traffickers, and illegal miners and loggers, assuming all maintain an armed capability to avoid prosecution and imprisonment. There is also the issue that there is often a strong crossover between poaching, narcotics and mining to support wider criminal or terrorist causes.

For what follows, we need to make a series of assumptions. The first is that the Government being protected is civilian and democratic in nature or an interim authority committed to establishing a civilian and democratic government in a timely fashion.

The second is that there is a legal framework for the detention or necessary killing of armed criminals or terrorists. Thirdly, this framework defines the enemy as criminals, not combatants, as per the Geneva Conventions; thus, those detained are not Prisoners of War.

There will always be real-world conditions that fail to correspond to the ideal used here, but this allows us to usefully simplify the training, equipment and organisation of military units to where insights can be gained. This is about military units, not police forces or paramilitaries or militias.

Competent armies should not require much additional training to contribute usefully to these operations, but some training is required or adaptations of training that are already well understood.

Lethal Force

Using lethal force to break the collective will to endure in combat defeats any type of armed opponent in any environment. Lethal force is the most effective and efficient method of breaking collective will. As Clausewitz made clear, killing does not set forth or resist the policy but rather removes the violent armed objectors that seek to counter it.

In general terms, killing the wrong people (civilians) may undermine the political objective being sought. Whether it does or not will be extremely context-specific to the policy and to whose policy is being set forth. How proportionately, precisely, or discriminately lethal force is applied will depend on the tactics employed. Thus, Rules of Engagement (ROE) are those limitations on lethal force and military activity that armed forces use to ensure that force does not undermine policy.

It can thus be reasoned that at the campaign level, the Army's mission should be to defeat the criminals or insurgents, as in breaking their collective will to endure in combat.

Defeating the enemy creates freedom of action to do everything else. It does not matter if Government policy ensures everyone has a red front door. An army's job is to kill or capture anyone who seeks to contest the colour of the front door violently. Non-violent opposition is everyday politics, not something the Army should worry about.

In 2005 The British Army's ADP Land Ops stated,

"Neutralizing the insurgent and in particular the leadership forms part of a successful COIN strategy. Methods include killing, capturing, demoralising and deterring insurgents, and promoting desertions. This is an area in which military forces can specialise and should be a focus for COIN training. The aim should be to defeat the insurgent on his own ground using as much force as is necessary, but no more."

Para 0156 e, ADP Land Ops AC 71819.[1]

This was useful guidance of the highest order. The later guidance, 2009 JDP3-40, stated, "Security cannot be achieved solely through the presence of military forces, or just by killing or capturing adversaries." This implicitly contradicted that which was stated as true in 2005. Yet Armies exist to use armed force against an armed force, so the primary mission of an Armed Force is to kill and capture. Security problems unrelated to armed criminals are not a concern of the Army.

Killing and capturing are important because lesser forms of operation aimed at "disrupting" or "dislocating" while useful, may allow the enemy to survive. Dead and captured cannot return at some later date to re-contest any issue they see fit. Warfare against irregular forces is won in the same way as regular forces. The only major difference is that force usually has to be employed far more precisely, discriminately and proportionately. This is because lethal force will be applied close to or within a population that you are politically / legally required to protect. This is national jurisdiction, and not the Law of Armed Conflict. The other difference is that lethal force will be focused at the individual level for the reasons already stated. This is a general distinction from fighting regular forces, where operations would seek to defeat units and formations in part or as a whole. It should be noted that irregular forces operating or based outside the relevant national borders; thus, jurisdiction can be subject to the same levels of force used against regular armed groups.

Training

More than any other type of operation, soldiers must be briefed on the cultural, social, and political realities of the situation they are entering as honestly and accurately as possible. They also must be carefully briefed and trained in when and when not to use lethal force and that if they kill someone under circumstances that seriously transgressed the conditions stated, they may be liable for murder.

The context of all security and counter-terrorist operations varies so much that it is impossible to prepare for any circumstance specifically, so a general framework of tactics, techniques and procedures needs to be built into annual training or adapted from more common techniques. It should also be noted that these skill sets can be applied within general war as and when required. In general terms, these are:

- Framework Patrols
- Observation Posts
- Ambushes
- Key Point Security
- Vehicle Check Points

- Cordons
- Public Order Training.

This represents the basis for a security training package, which could be delivered within 14 days, assuming that the soldiers concerned are already competent in all the field skills a normal infantry battalion should have. There is the additional consideration that security operations can be so manpower-intensive that soldiers from other arms, such as Cavalry, Artillery, Air Defence and Engineers, may have to be trained to accomplish some of these tasks, such as Key Point Security and Vehicle Checkpoints.

Overall, the aim of the Army in Security operations is to reduce the enemy's freedom of action to the degree that his capacity for action is greatly degraded because he fears being detained or killed during the conduct of a criminal or terrorist act.

The more decisive activity where intelligence is used to mount specific operations against known individuals or locations is usually the domain of Special Forces or Specialist Counterterrorism units and falls outside the scope of this work.

Framework Patrols

Regardless of doctrinal definitions, "Framework Patrols" essentially serve two purposes: to demonstrate freedom of action and to deny the enemy freedom to conduct any activity.

Patrols must have the legal authority to move across the ground and within any property, although that authority must be exercised sensibly in terms of cultural or social sensitivities, and junior commanders need to be selected and trained with that level of judgement.

Combined with that, they must be able to stop and search any person or vehicle to provide the necessary deterrent action to make the enemy feel unsafe or hunted. The training for such patrols, whether mounted or dismounted, must emphasise that no patrol ever works in isolation from another and that competent patrol action will result in unpredictable patrols when and where they appear and will always operate in mutual support in case one patrol comes under attack. There is, of course, a high likelihood of the enemy using sniper attacks and improvised explosive devices to attack patrols.

Hence, training must avoid setting patterns or routines that enable the enemy to plan and execute such attacks. This brings us to the issue of observation posts.

Observation Posts

Observation posts can be both overt and covert. Both should work together to suppress and restrict enemy movement or activity and support intelligence gathering. Overt observation can be as varied as observation towers equipped with ground surveillance radars and powerful electro-optics, to UAS or airborne observation. Ideally the enemy should not know when or of they are being watched. Modern technology has provided various remote sensors, such as covert cameras using face recognition software. Still, their existence does not remove the necessity for traditional manned cover surveillance from well-hidden observation posts, especially when an immediate lethal reaction to enemy contact is required. The capability to place four well-trained soldiers inside any environment and have them maintain observation and communication for 5-6 days without re-supply should be nothing unusual for infantry soldiers, even given the required high levels of discipline and personal administration. 10-14 days are normally only the domain of Special Forces. Covert Observation Posts can be placed almost anywhere in rural terrain or farmland, and in urban environments, have even operated from within civilian-occupied buildings without the homeowners being aware. The insight here is that while the skills needed to conduct covert observation should be inherent to normal infantry training as in defensive and screening operations, the security and counter-terrorist operations will place soldiers in greater proximity to a civilian population from which they still need to avoid detection and need to train for circumstances where a soft compromise may occur where civilians are alerted to their presence either by accident or design. The enemy can use women and children with or without dogs to run counter-surveillance plans to detect security force observation.

Ambushes

Ambushes were a common activity during the British anti-terrorist campaigns in Malaya, Kenya and Cyprus but were only appropriate to the enemy moving on known routes in any number. The classical ambush has largely been replaced operationally by the "Observation Post React", (OP React) as in an observation post designed to apply lethal force against an identified and armed enemy. There is nothing to suggest that an area ambush cannot be accomplished by employing several "OP React" designed to trap the enemy into a killing area and prevent their escape or rescue by others. The circumstances are many and varied, but the training for this can be usefully supplied by giving Observation Post training an additional context, which reduces training time and resources.

Vehicle Check Points

As with observation posts, Vehicle Checkpoints (VCPs) come in two basic forms with some specific variations. VCPs are either permanent or "snap". Permanent VCPs operate overtly, subjecting vehicles and the people within them to varying degrees of scrutiny, including technical means such as retina scans and facial recognition. In many circumstances, such checkpoints will have to allow for large numbers of civilians moving on foot, so the traffic volume is a substantial planning factor to ensure everyday civilian life is not too severely impacted.

Snap VCPs are launched at unpredictable times and places by foot patrols, which can even be dropped by helicopter onto remote roads and trails to interdict either specific vehicles or groups. Snap VCPs offer a powerful tool within the wider framework of operations that makes the enemy's chance of moving weapons and equipment or avoiding detention substantially more difficult than if such measures were not in place.

The legal and practical constraints of bringing civilian vehicles, including large trucks, to a safe halt within a road system are not simple and require substantial instruction. The use of lethal force to bring a vehicle to a halt is not something more democratic powers would allow unless there was a clear risk to life as a result of the vehicle failing to stop. The existence of vehicle-borne or suicide IEDs substantially complicates this as it does all other activities. Suicide bombers or suicide mission actions are not something unique to the Middle East and were common in Sri Lanka and Southeast Asia long before the "Global War on Terror", so it should not have been a surprise to anyone. While contexts will vary, training must account for this across all the skill sets and TTPs.

Key Point Security

Key Point security is normally a platoon or sub-unit task aimed at securing key infrastructure that the enemy may target for whatever political or military advantage.

Power Stations, hydroelectric dams, mining facilities and airports are just a few of almost endless examples. The main effort of any key point security activity should be anything that complicates the enemy's ability to gather information necessary to planning any attack or disruption, such as variations in routine, security post locations and numbers present. Within the facility, the force generally needs to split into enough sub-divisions to account for a guard force, a quick reaction force, support and administration, and rest periods. Static guarding should not be a boring and mundane activity likely to sap morale and create an unenthusiastic or passive attitude to the task.

This is sometimes unavoidable, but the sensible use of practice alerts, satellite patrolling and short-term ambushes on possible approach routes or stand-off attack positions (called Base plate checks in the British Army) should all contribute to a sense of purpose and dynamism.

Cordons

Cordons seek to do two things: first, prevent access into or out of the area and second, secure an area to the degree that an activity can be conducted within it with greater assurance than if that were not the case. Activities within the cordon could be arrests, searches of locations or making safe suspected explosive devices. A cordon might also be needed after a mass casualty incident or terrorist attack.

Cordons are not as simple as many might suppose. Manpower will often be in short supply, so the cordon may not be as coherent as optimally desired; plus, it may extend for a period far longer than imagined and might be conducted in bad weather, be that cold and wet or hot and dry. There will be the need for a quick reaction force to respond to any incidents on the cordon and to rotate soldiers on and off for feeding and rest. The cordon may also have to adapt to changing circumstances, meaning it must be repositioned in part or whole while not losing effectiveness. A cordon planned for 30 mins may run to 24 hours, so training needs to account for that.

Public Order Training

Public Order Training more commonly known as Riot Control, is a skill set that spans security operations to any occasion where widespread public order may be a problem, such as refugees or large numbers of detained persons. Equipment such as shields, shin pads, visors and batons are easily stored and moved and can be distributed quickly. Still, training is essential to prevent serious injuries or unnecessary deaths to unarmed civilians. Soldiers tend to be fitter and stronger than policemen and can react with a high level of physical violence, so this needs high degrees of discipline and restraint. Other riot control issues, such as chemical irritants, sound generators, and water cannons, will have to fall within the legal framework of the training and thus are subject to the relevant government policy. As a general approach, soldiers must avoid situations where they are pelted with missiles or have their freedom of manoeuvre contested. Controlling violent crowds requires the infliction of harm and detention to break the will to persist in disorder.

Specialist Security Units and Personnel

In some areas, security and counter-terrorist operations require specialist units and personnel. Military Working Dogs would be one example. Counter-IED would be another. By far, the most decisive specialisations are found inside intelligence, which is the key to bringing the enemy into contact with the security forces under the most advantageous circumstances. Almost every aspect of security and counter-terrorist intelligence operations is highly context-specific. Therefore, any competent army must maintain a broad array of intelligence specialisations and the capacity to interface and work with foreign intelligence agencies and local police forces when and where appropriate. Regarding the maintenance and support of such units and skill sets, most governments should recognise that training other armed forces to develop their capabilities provides significant diplomatic value if correctly applied.

Competent armies do not have to choose between warfighting, security, and counter-terrorist operations, except when the priority of operations overrules all else.

Colombia and Mexico are examples of this. If warfighting is the default setting, then leveraging that to produce a security and counter-terrorist capability is substantially easier than the alternative. For example, it is trivially simple to retrain infantry battalion Mortar and Anti-tank platoons to the security skills but expensive and time-consuming to create Mortar and Anti-tank platoons from scratch given a sudden need. The same is true across almost all arms and services. You can make a good soldier into an adequate policeman more easily than a good policeman into an adequate soldier.

Endnote

1 Horribly undermined by later publications.

18

OBSERVATIONS CONCLUSION

OK, so you have waded through the book. So what?

Conclusion chapters often look like an After-Action Report, where the preferred version of events can be committed to paper rather than the actual truth of what transpired. Hopefully, we can avoid this, and the reader can agree that the core parts of the work were:

a. You cannot predict why, when, against whom or where the next war will be, so be prepared for how to fight "a War" is optimal.
b. There is no point in having an Army or Land Force you cannot afford to equip, train, and organise. The size of that land force is a political/economic problem, not a military one.
c. A deep understanding of ideas is the bedrock of knowing how to equip, train, and organise within a budget while being militarily capable.

There is a phrase in Hebrew that approximately translates to the shortest path to ridicule is prophecy. Anyone claiming to tell you what tomorrow's wars look like is playing with lives. In sharp contrast, what this work aimed to tell you was to prepare for what we know today. If you are prepared for every day as it comes, then the future holds fewer surprises than if you chewed and debated on what the future may look like. Everything this work discussed was achievable and mature at the time of publishing or merely required trivial amounts of time and investment to make a reality. The correctly trained, equipped and organised army you have parked on the square this evening is the one that will still be correctly trained, equipped and organised a year from now. Thus, understanding what manifests the most likely correctly trained, equipped and organised land force is a constant and iterative process requiring annual review and gradual evolution.

It is likewise completely pointless to discuss tactical action in terms of function unless it is made real by unit organisation, training, and equipment. Hopefully, this work has clarified that training, equipment, and organisation directly determine how you fight and operate.

An oft-used phrase among writers is, "I learned about this by writing about it." Such statements may be true, but writing alone can never create the true level of expertise required to solve problems. Works without actionable solutions often justify their statements with "the need to start the conversation." In contrast, my baseline for this work was whether I could explain, teach, and demonstrate the organisation, skill set, or activity I was referring to. Even so, that may sometimes fall short of being a solution. If others have better work, this should spur them to present it. Every idea advanced in this work should be legitimately challenged with the words "Show me how that is done" and "How does this work." "Where does this idea come from?" should also be something a sceptic should ask.

Despite the details of the discussion, this was not a cookbook. Nowhere does this work contain recipes for success. If you understand Cost, Weight and Complexity as the conceptual framework that best informs equipment, then "Price, Politics, and People" is the harsh truth of force development. Suppose the wife of the current Prime Minister is Colonel in Chief of the Women's Auxiliary Balloon Corps. In that case, you probably won't see them disbanded and their budget allocated to something better.

I hope that any positive outcome from this book would be a reader who can now do something better professionally than they would have had they not read it. This is a somewhat utopian ideal, as understanding does not always lead to action, but insight and realisations are rarely bad.

The less positive but extremely useful outcome might be an increased scepticism of much of the writing about war, warfare, and land forces that do not account for the practical challenges that impact the ivory tower ideas untested against legitimately sceptical professional audiences. Nothing aids conceptual and practical development better than a high degree of rigour. Social media will always attract a hostile and mostly uninformed audience via text rather than practical knowledge; thus, the art and science of modern military thought have seen countless nugatory debates that should not and cannot be translated into training, equipment, and organisation and, thus, are mostly devoid of insight for those who may have to fight the wars.

The limit of this work is that all that has been written needs to be tested in written orders, training programs, and detailed equipment and loading plans for units. Suppose you cannot or will not do that. In that case, the debates and arguments flounder in the pages of military journals or, worse, social media posts written by those of unknown provenance which can ape professional language.

Conducting modern warfare is an entirely practical skill. Theory, concepts, and doctrine are all required, and this work strongly advocates for an approach based on simple and rigorously tested ideas. Bad theories, concepts, and doctrines are exposed through practice and material decisions in the real world, ideally before firing rounds in anger.

Hopefully, this book has proven or indicated that a well-trained, equipped and organised land force is not some unachievable and vastly expensive ideal but something most nations can logically aspire to within current resources and cognisant of political and economic reality.

Finally, it may be useful to condense the total of what has gone before in 17 propositions:

Proposition 1: You cannot tell where, when, why or against whom you will next fight. You must know how to fight regardless of that information.

Proposition 2: The size of an army is a strictly political choice based on budget and priorities. How much a nation spends depends on how frightened or relevant it wants to be.

Proposition 3: There are few overall non-discretionary choices in a land force. Equipment, training and organisation are all choices governed by cost versus value. You can have an effective army without tanks.

Proposition 4: If the tank is obsolete, so is every armoured vehicle in every army.

Proposition 5: Signature may be more important than armoured protection.

Proposition 6: Training best defines an army's effectiveness in combat, regardless of size.

Proposition 7: All Divisions and Brigades/Formations are assemblies of units. The unit organisation defines the choices of training, equipment and organisation.

Proposition 8: Limiting unit organisation to 100 vehicles per unit (and other defined parameters) forces hard choices, thus, insights into where the investment balance should lie.

Proposition 9: The communications used to control, manoeuvre, and sustain fires should form the C3I backbone for all levels of command below the Corps/Field Army.

Proposition 10: The most important equipment choice a modern army must make is the communications systems it uses to support and sustain C3I.

Proposition 11: Effective infantry is best enabled by well-trained junior leaders, platoon commanders, and company commanders who can make quick and effective decisions while fighting and operating with any number of men and any scale of equipment.

Proposition 12: All things being equal infantry section, platoon and company effectiveness is more strongly governed by communications, sensors and access to supporting fires than by light weapons and platoon or section organisation.

Proposition 13: Only badly trained or badly led soldiers are overloaded.

Proposition 14: The best purpose of reconnaissance is to find where the enemy is and where he is not so that he can be engaged under the best possible circumstances by the most effective means.

Proposition 15: Killing drones, cyber, air defence and EW operators should have an equal focus on destroying their equipment.

Proposition 16: The effectiveness of Fires is best understood from its potential to break the human will to endure or persist in combat. Break people, not equipment.

Proposition 17: The test of an army is its readiness for war.

This is just the highlights of what we have discussed. Other insights may be more apparent to the reader than they are to me. If that is the case, I think this is the best I can hope for.

INDEX

www.ingramcontent.com/pod-product-compliance
Ingram Content Group UK Ltd.
Pitfield, Milton Keynes, MK11 3LW, UK
UKHW052103030125
3948UKWH00049B/473